XHTML/CSS BASICS FOR WEB WRITERS

MARGARET W. BATSCHELET
University of Texas at San Antonio

PEARSON

Prentice
Hall

Upper Saddle River, New Jersey
Columbus, Ohio

D1312439

Library of Congress Cataloging-in-Publication Data

Batschelet, Margaret.
    XHTML/CSS basics for Web writers / Margaret Batschelet.—1st ed.
        p.   cm.
    ISBN 0-13-172014-7
    1. Web site development.   2. XHTML (Document markup language)   3. Cascading
style sheets.   I. Title.
    TK5105.888.B379 2007
    006.7'4—dc22

                                                                    2006017747

Editor in Chief: Vernon Anthony
Senior Acquisitions Editor: Gary Bauer
Editorial Assistant: Daniel Trudden
Production Editor: Louise N. Sette
Production Supervision: Integra
Design Coordinator: Diane Ernsberger
Cover Designer: Terry Rohrbach
Production Manager: Pat Tonneman
Marketing Manager: Leigh Ann Sims

This book was set by Integra. It was printed and bound by R.R. Donnelley & Sons Company.
The cover was printed by R.R. Donnelley & Sons Company.

Pearson Education Ltd.   Pearson Education Australia Pty. Limited
Pearson Education Singapore Pte. Ltd.   Pearson Education North Asia Ltd.
Pearson Education Canada, Ltd.   Pearson Educación de Mexico, S.A. de C.V.
Pearson Education—Japan   Pearson Education Malaysia Pte. Ltd.

10 9 8 7 6 5 4 3 2 1
ISBN 0-13-172014-7

BRIEF CONTENTS

CONTENTS

3   CASCADING STYLE SHEET BASICS   37

4   WORKING WITH TEXT: HTML TAGS   53

7   WORKING WITH COLORS AND BACKGROUNDS   109

8   LINKING PAGES   124

9   WEB SITE DESIGN   149

12   USING MULTIMEDIA   212

PREFACE

I wrote *XHTML/CSS Basics for Web Writers* because I wanted my students to have a text that would explain code in a straightforward, easy-to-follow format, while also providing information about creating effective design and text. Textbooks on Web writing frequently skip over HTML and Cascading Style Sheets (CSS) as if the topics aren't really necessary for Web page creation, and books on HTML and CSS pay little attention to the basics of page design and effective Web text. Moreover, some earlier textbooks on Web writing that did cover HTML (like my own *Web Writing/Web Designing*) taught the HTML of the late nineties rather than standards-compliant XHTML and CSS.

XHTML/CSS Basics for Web Writers is a hands-on textbook for Web page design and construction. It takes students through the Web design process, from assembling and inventorying information for a Web site through writing the text, locating graphics, and creating a page design. However, the principal focus of the text is using Web code, specifically XHTML and Cascading Style Sheets. My goal in creating this text was to provide students with a guide to the process of Web writing as well as instructions in using specific coding techniques. The text is intended for students with varying levels of experience in creating Web sites, from absolute beginners to those who need to update their skills for standards-compliant design. Although most of my own students are communication or English majors, *XHTML/CSS Basics for Web Writers* is not keyed to any particular discipline. In fact, these days Web sites and Web site construction are so widespread that students in any program may find the text helpful.

Organizing a book on Web writing is always difficult because the process of writing and designing a Web site is frequently nonlinear. I've tried to make the content of each chapter as transparent as possible so that instructors and students can follow the order that works best for them.

The text begins with a chapter on preliminary steps in creating a Web site, including choosing a topic (for assigned Web sites), analyzing purpose and audience, assembling information, and creating a preliminary organization. Although there are clearly preparatory steps in the Web design process, the process itself is seldom linear. Thus, the first chapter on code, chapter 2, "HTML Basics," provides information for creating a basic page, including text tags, the basic image tag, and a simple link tag. All of these topics are covered in greater depth in later chapters, but these first two chapters allow students to develop a functional Web page from the outset. Similarly, chapter 3, "Cascading Style Sheet Basics," describes the way CSS works and provides the basic style properties necessary for developing a simple, text-based page.

The following chapters provide more detail about both XHTML tags and CSS properties. Chapters 4 and 5 describe tags and properties used with text, as well as ways to make Web text more readable. Chapter 6 covers the use of images on the Web, both locating images and placing them on Web pages. Chapter 7 provides information about systems for designating foreground and background colors, as well as concerns regarding color legibility and accessibility for color-blind users. Chapter 8 covers links, including writing tags for relative and absolute links, choosing link colors and styles, and creating effective link rhetoric.

Chapters 9 to 11 cover the most challenging part of Web site creation: design. Chapter 9 provides basic information about page design, including design principles, Web site accessibility, and widespread design errors. Chapter 10 covers beginning CSS design principles: the div and span tags and the float property, as well as the CSS box model. Chapter 11 provides instruction on positioning, both absolute and relative, as well as information on using the z-index and the display property. The chapter closes with two examples of CSS design for navigation menus.

Chapter 12 provides information on using sound and video files, including coverage of the currently chaotic state of code for inserting these files into Web pages. This chapter also provides a brief overview of Flash animation. Finally, chapter 13 is an overview introduction to XML, the code that is becoming the backbone of the Web.

XHTML/CSS Basics for Web Writers ends with a series of appendixes covering topics that may be of use to individual students and classes. Appendix 1 covers HTML table tags. Although tables are no longer used for page design within standards-compliant pages, they are still used for presentation of data, and CSS provides some additional properties for table design. Appendix 2 presents some techniques for image replacement, a method of inserting graphics in the background of pages while providing a text equivalent for nongraphic browsers. Appendix 3 provides a method of creating dual style sheets, one for print and one for the screen. Appendix 4 presents an overview of blogging. Appendix 5 provides a table of the standard character entities since XHTML requires that all characters be encoded. Finally, appendix 6 includes a list of deprecated and browser-specific tags and attributes so that students can make sure they're using standards-compliant code.

In *XHTML/CSS Basics for Web Writers* I've tried to provide both instructions and background information that will enable students to create effective, standards-compliant Web sites. I hope that students who use this text will find the process of Web design both challenging and rewarding, as I do.

ADDITIONAL RESOURCES

For the Student

Go to www.prenhall.com/batschelet to access links to online examples, templates, and downloadable resources.

For the Instructor

This textbook is supported by an Instructor's Manual with Test Item File and a PowerPoint Lecture Presentation package. To access supplementary materials online, instructors need to request an instructor access code. Go to www.prenhall.com, click the Instructor Resource Center link, and then click **Register Today** for an instructor access code. Within 48 hours after registering you will receive a confirming e-mail including an instructor access code. Once you have received your code, go to the site and log on for full instructions on downloading the materials you wish to use.

ACKNOWLEDGMENTS

I must first of all acknowledge the help I received from my students in Communication 3413: Writing for New Media. My students used versions of these chapters in our class; and through their feedback and my own observations of both successes and failures, I was able to revise the material, making it more useful and, I hope, approachable. I would also like to acknowledge the reviewers of this text: Randy Brooks, Millikin University; Huatong Sun, Grand Valley State University; Erin Karper, Niagara University; Steven Krause, Eastern Michigan University; Jeffrey Jablonski, University of Nevada Las Vegas; Roger Grice, Rensselaer Polytechnic Institute; and Ellen Taricani, Pennsylvania State University. In addition, I am grateful for the encouragement I received from my colleagues in the Communication Department at the University of Texas at San Antonio. Finally, as always, I am thankful for the support of my family—Bill, Josh, and especially Ben, my very own tech advisor.

Web Design Preliminaries

Chapter 1

WEB WRITING

In many ways Web writing has little in common with writing as we usually think of it. Writing a Web page may include writing text, of course, but it may also include elements that are seldom part of academic writing. To be a good Web writer, it's not enough to be good with words: you need to know something about graphic design, about the use of images, about color combinations, about typography, and, most of all, about using Web code. In fact, it may be best to think of the process you use with a Web page as designing rather than writing. Your text will be a part of what you do, but only a part. The rest will involve a variety of skills and materials that may be unfamiliar to you at this point.

The major focus of this book is using Web code: HTML/XHTML and Cascading Style Sheets (CSS). However, there are several preliminary steps in designing a Web site that you should undertake before beginning to create individual pages. In this chapter we'll discuss some of these preliminaries, including deciding on your purpose and audience, examining your competition in terms of Web sites that are similar to yours, and developing and organizing possible content.

WEB SITE TOPICS

If you're using this book in an academic setting, you may have been assigned to create a Web site; your instructor may have already given you guidelines for a topic, or the choice may be up to you. The standard advice given to students choosing paper topics can also apply to Web sites: choose something that appeals to you personally. Because Web sites involve a significant amount of effort, you'll need a topic that you won't mind working with over several weeks. If you're given some flexibility in the kind of content you can use, you should also look for topics that can be presented using both text and graphics. As you probably know from your own experience with the Web, successful

---

EXERCISE: CHOOSE A TOPIC

If you haven't already chosen a topic for your Web site, try brainstorming for one now. You might consider some of these general possibilities:

- ❑ A theory or an issue in your major area of study or a concept in your major that people have difficulty understanding
- ❑ An author, director, or artist and his or her major works (e.g., Martin Scorsese and his films, Sarah Smith and her novels)
- ❑ A social or political issue or problem
- ❑ A technique or set of skills for which people need instructions and/or explanations of materials
- ❑ A location (e.g., city or park) or historical site
- ❑ An organization or nonprofit agency
- ❑ A commercial site for a product(s)

Web sites usually involve a variety of materials. Finally, choose a topic that's worth your time; you don't want something so lightweight that you'll grow tired of it easily.

THE WEB DESIGN PROCESS

Like any other project, Web sites require advance planning to be successful. It's always easier to consider questions of audience, purpose, content, and organization before you commit yourself to a particular design. The old adage still holds true: the more time you put in at the front end of a project, the less time you'll have to put in at the back end.

Purpose and Audience

As you plan your site, your first questions should concern the purpose for which the site is being created and the people most likely to use the site. For example, if you were creating a site for a campus organization, one purpose might be to attract new members. But another purpose might be to communicate with current members who need information about the club's activities. Still another purpose might be to create a site so that your organization could be listed among the other linked organizations on the university's Web site. In each case you'll be considering a slightly different audience with slightly different needs: Potential members would need to know what the organization's purpose is, what type of activities the organization sponsors, when, where, and how often it meets. Current members might be more interested in information about current club activities and schedules. And university administrators would want to make sure that the club was appropriate to be listed on the university site.

DESIGN STEP 1: DEFINE YOUR PURPOSE

Assuming you've already decided on a topic for your Web site, try defining its purpose in one or two paragraphs. You might consider these questions:

- ❑ What overall purpose will your site fulfill?
- ❑ What will your site enable visitors to see, understand, or experience?
- ❑ What need will your site meet that isn't currently being met by other sites?
- ❑ What are the three most important goals for your site?

You can begin by listing the goals you have for your site: what do you want the site to accomplish? If your site is for an organization, Louis Rosenfeld and Peter Morville (1998) suggest that you begin by considering the organization's mission and the ways your Web site will support it. You can even compose a mission statement for the Web site to help focus on how the goals of the site and the goals of the organization might mesh. Doing this can help you see your site in relation to the other parts of the organization. However, whether you create a complete mission statement or simply a list of goals for your Web site, you should spend some time thinking about what you want the site to do before considering how you want it to look.

Audience is closely allied to purpose—knowing whom you expect to come to your site will make a difference in what you want your site to accomplish. Moreover, the nature of your audience will affect the kind of organization, design, and navigation your site will use. Rosenfeld and Morville (1998) suggest brainstorming about not only the most important audiences for your site, but also those audiences you're less likely to remember—for example, members of other campus organizations who might want to work with your organization or members of the organization's national chapter who might want to keep up with what the local chapters are doing. It also helps to know how people are currently getting the information they'd be seeking on your Web site: from a newsletter, from printed brochures, from a recorded phone message, and so on.

Once you have a list of your possible audiences, you can rank them in importance and list the most important needs for each audience. You'll definitely want to meet the needs of your primary audience (the one that includes the majority of your visitors), but you should also keep the needs of the secondary audiences in mind as well—it's possible that you can meet the needs of your secondary audiences at the same time that you're meeting the needs of the primary one. If you can talk to real members of your potential audience (e.g., current club members or people who might be interested in joining your club), they can help you get some ideas about the types of information that you might include on your site. Ask them what they'd like to see on a site for your organization; ask them for names of Web sites that they use frequently, and try to define what they like and dislike about them; and ask them to suggest sites that might be similar to the one you're designing that they'd recommend as models.

DESIGN STEP 2: ANALYZE YOUR AUDIENCE

Try to list potential audiences for your Web site, both primary and secondary. As you analyze your audience, consider these questions:

- ❑ Who is the primary audience for this Web site (e.g., people already knowledgeable about your subject or novices, current or potential members of the organization)?
- ❑ What do you want this audience to be able to do after they visit your site?
- ❑ What do you want this audience to think after they visit this site (e.g., should they want to find more information on the subject, should they be upset or concerned about an issue, should they be thinking about joining your organization)?
- ❑ How knowledgeable is your audience likely to be about Web site conventions (i.e., navigation, searching, and so on)?

Other Web Sites

After you've identified your purpose and audience, it's a good idea to see what kinds of sites already exist on the Web that are similar to the site you're planning. Doing this can give you an idea of what works and what doesn't for a site of this type. You're not trying to copy someone else's designs; you're trying to anticipate problems and solutions that you may encounter in constructing your own site.

You can also get ideas for possible content to include on your site or features to avoid. Seeing someone else struggle to master a particular aspect of design or content may help you avoid making similar mistakes in your own design. At the very least, it's a good idea to get some idea of your competition, the other sites out there that might attract the same kind of users you'll be trying to attract. Thus, for our campus organization site, you'd look at the other club Web sites already posted at your university and possibly at sites for other chapters of your organization on other campuses.

DESIGN STEP 3: STUDY YOUR COMPETITION

Try to find two or three sites that cover topics that are similar to yours. Go to a search engine and type in some keywords relating to your topic to see what comes up. Once you've located some sites to study, you can consider these questions:

- ❑ What's your first impression of the site, based on its look and feel?
- ❑ Is the navigation clear; is it easy to get around?
- ❑ How easy is it to find individual items?
- ❑ How effectively is the site organized? Is it segmented into logical, predictable divisions?
- ❑ What features and content do you particularly like about this site? What do you dislike about it?
- ❑ Are there any features you'd like to transfer from this site to your own site?

Content

At this point in the process—now that you have some idea of the reason your site is being built, the people who are likely to come to it, and the features built into other, similar sites—you can begin to think about the actual content of your own site. Although you probably won't anticipate everything that should be part of the site at this point, you should have a clear idea of the majority of the topics you'll want to cover—you don't want to end up adding a lot of pages at the last minute! You'll probably have some content in mind from your discussions with potential audience members and your examination of other sites.

You can begin by listing possible content. If you're working with an organization, you can ask other people within the group to list topics they'd like to find on the new site. Or you can talk to other people involved in the project to brainstorm ideas for topics to cover. Once you have input from everyone involved, you can make a list of probable topics for the site, combining items that seem to go together and keeping track of topics that are mentioned frequently. Clearly, not everything people suggest can be included; some topics will probably fall into the more-trouble-than-they're-worth category, and some will reflect specialized interests that aren't shared by other potential users. But many of the items may show up on several lists, reflecting a consensus of interest among your potential site visitors. Create a master list of the topics that seem to be most relevant and most widely requested. Then, rank the items on the list in terms of their importance for purpose and audience. Again, ask for input from other group members regarding their own ranking of the topics.

Once you have a list of the content you think you'll include, Patrick Lynch and Sarah Horton (2001) recommend creating a "content inventory": that is, a list of existing content as well as content that you'll need to create. The existing content probably won't be in the form of Web pages, but you may find other types of information that are available—brochures, reports, newsletters, and other documents that you'll want to include in some version on the site.

Along with these already-existing documents, there will undoubtedly be some information that will have to be created from scratch. You can devote more time to areas in which you need to develop content than to areas in which the content already exists and is ready to be placed onto a Web page template. In fact, you'll probably spend the bulk of your time developing and placing content from this point on; that's frequently the most time-consuming task for a Web site.

DESIGN STEP 4: LIST PROBABLE CONTENT

Take stock of the content you'd like to include on your Web site. If you haven't already done so, try listing possible topics you'll want to include. If there are other people involved in your site, have them list possible content, too; then try comparing your lists to see what topics you all agree on. Try to come up with a list of five to ten topics that will be included in your site.

DESIGN STEP 5: INVENTORY YOUR CONTENT

At this point, before you actually begin working on your Web site, make a list of all the content you already have available: text that you've written, pictures you want to use, perhaps even audio or video files. Once you know what you have, you can decide what you need to locate before you begin putting your Web pages together.

Once you have a fair idea of the amount of content you'll need to include on the site, you can begin organizing the site to fit the content.

ORGANIZING INFORMATION

The way you choose to organize your information will depend on both the purpose of your site and the nature of the information itself. Information architects Louis Rosenfeld and Peter Morville (1998) suggest approaching Web site organization by looking at two different levels: organization schemes and organization structures. Organization schemes "define the shared characteristics of content items and influence the logical groupings of those items" (26). Organizational structures "define the types of relationships between content items and groups" (26).

Organization Schemes

Organizational schemes can be exact or ambiguous. Exact schemes include familiar ordering principles such as the alphabet, chronology, or geography; they can work well if users know exactly what they're looking for. If you want the e-mail address for an officer of the club, for example, it helps to have an e-mail directory organized alphabetically rather than on some other basis. Exact schemes are fairly easy to design and maintain since the categories are very straightforward and easy to use. Thus, if the category is "all last names beginning with R," it will be fairly easy to figure out what content goes into that category.

However, not all information can fit easily into exact categories. Ambiguous organization schemes use categories that you may not be able to define exactly but that may be more useful for the kind of content you're working with. You might organize your content by topic, for example, like the yellow pages in a phone book. This might not be the only organization that you use, but it could be extremely useful to have topical organization available. With topical organization your classification system should be made up of categories and subcategories that clearly define the type of content that they include and exclude. Just be sure that your topics cover not only the current information to be included on your site, but also future information that may be added later.

Another type of ambiguous scheme is audience specific. For example, many university Web sites are divided into sections for current students, future students, parents, faculty and staff, and alumni, with tables of contents listing

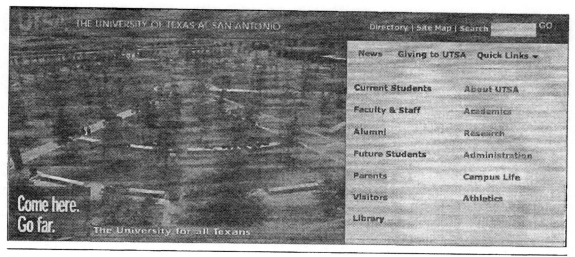

FIGURE 1–1 Two organization schemes

information of interest to each group. Even though these sites are organized for particular visitors, visitors can usually also visit the sections of the site that are organized for other groups.

It's possible to have more than one organization scheme for a site. For example, my university's Web site has both audience-specific menus and topical menus, giving visitors a choice of ways in which to approach the site's content (see Figure 1–1).

Here visitors can either choose menus for Current Students, Faculty & Staff, Alumni, Future Students, and so on, or they can choose topical menus under headings like About UTSA, Academics, and Research.

If you choose to use multiple schemes, however, be careful not to start blending elements into common menus and a common organization. Each organization scheme should maintain its own identity and its own menu. A mixed-up organization makes it difficult for your users to get a mental image of the way the site is organized, and they may become frustrated when they try to find a particular topic in the blended organization.

Organization Structures

The organizational structure of your site will depend on the organization scheme you're using and will define the ways users can navigate your information. If you decide to use a chronological organization scheme, for example, you may end up with a somewhat linear organizational structure so that users can follow the chronology step by step. Two of the most familiar organizational structures for Web sites, however, are nonlinear: hierarchical structures and hypertext structures.

Hierarchical Organization. **Hierarchical organization** provides a structure that will be familiar to users and designers alike. Top-down hierarchical structures with branching nodes, mutually exclusive subdivisions, and parent-child relationships are well known from outlines, organizational charts, family trees, and even the typical organization of textbooks like this one. Thus, users will understand quickly how the organization works, which will help them move easily among the various levels. One of the principal questions you'll face with a hierarchical organization, however, is what Rosenfeld and Morville (1998) call "the balance between breadth and depth in your information hierarchy" (38). Breadth refers to the number of options users have at various levels; depth refers to the number of levels the site has. If a site has too much depth, users may have to click through a bewildering number of levels to reach the information they need. If a site has too much breadth, users may have too many options on a menu and too little content on individual pages. You'll need to find a balance between breadth and depth, limiting the number of options on your main menus and being careful not to make users click through more than two or three levels to find what they need. Rosenfeld and Morville (1998) suggest that new Web sites that may add content later on should aim for a structure that is "broad and shallow."

A hierarchical organization of information for a campus organization might look like Figure 1–2. Here the hierarchy implies that the organization's mission and purpose represent what the organization is all about. Thus, the organization's activities and history will grow out of its mission and purpose, whereas its calendar relates to its activities, and its special programs relate to its history.

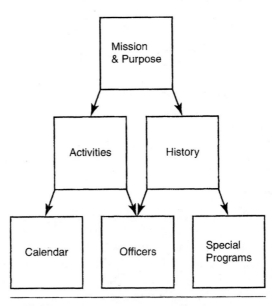

FIGURE 1–2 Hierarchical organization

The officers, involved in both the activities and the history of the organization, are linked to both topics.

Hypertext Organization. **Hypertext organization** is nonlinear organization that concentrates on chunks of information and the links between them. Hypermedia systems can be made up of a variety of files, including text, graphics, video, audio, and animation. These segments can be organized hierarchically, nonhierarchically, or both, through a flexible system of relationships. However, because links are frequently based on relationships that are personal to the designer, some users may become confused or lost while clicking through a hypertext system. Rosenfeld and Morville (1998) suggest that hypertext organization should be used "to complement structures based upon the hierarchical or database models" (40). After first designing a hierarchical structure for the information, the designer can "identify ways in which hypertext can complement the hierarchy" (40).

A hypertext organization of information for a campus organization might look like Figure 1–3. Once again, the mission and purpose of the organization is central to the site (and would probably be the entry page). However, the paths leading from that central page provide a variety of possibilities: The activities page links to videos and pictures that are (ideally) related to those activities. The officers page also links to the pictures page so that visitors can see pictures of the officers. Special programs link to relevant video files, and so on. As you can see from this example, hypertext organization is much more wide open than hierarchical organization.

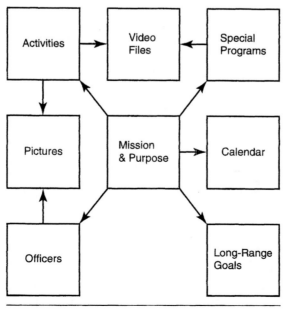

FIGURE 1–3 Hypertext organization

Working with Content

Deciding on the organization scheme and structure you want to use will involve looking at both the information you'll be including and the way you want an audience to approach it. For example, how many choices do you want to present to your users on your main screens? You can consider what natural relationships may exist between your information groups. Jennifer Fleming (1998) describes a method for doing this called "content chunking," which is used by some information design firms. With this approach you write your content items on index cards, one item per card. Then you try to organize the cards into piles of related items, assigning a label to each pile. It helps to have other people in your organization try the same exercise, making their own piles and labels. When you've finished, you can compare the different ways of grouping and labeling, analyze the results, and come up with the structure that fits your information most effectively, using the piles as groups of topics and the labels as the higher-level classification for each group.

For example, returning to the Web site for a campus organization, you might have content about the organization's history, both local and national; you might have profiles of current officers and faculty sponsors; you might have descriptions of past activities the organization has sponsored and descriptions of upcoming activities that the organization is planning. You'd probably have basic information such as contact information, meeting times and places, dues, and any requirements for membership, as well as a calendar for the coming year. You might have formal documents like the organization's charter and perhaps the campus organizational guidelines. You might even have a Frequently Asked Questions page. To organize this information, you'd create an index card for each topic and then try possible groupings. For example, you might group history together with profiles of officers in a who-we-are group. Information about past events and planned future events could go along with the calendar to create a what-we-do group. Contact information and meeting times could be grouped under "how to find us," and dues and applications could be grouped under "how to join."

Another person could look at these topics and find different ways of organizing them and different labels for the groups that are established. Because there are frequently a number of effective ways to organize information, it's a good idea to get as much input from others as you can. Perhaps an alternate organization will

DESIGN STEP 6: ORGANIZING YOUR CONTENT

Try your own version of Fleming's content chunking. Go back to the list of topics you want to cover in your site, and write each of them on an index card or a slip of paper. Then try grouping the cards, putting them into piles based on commonalities between the topics. Once you have your topics organized, try to write a label for each of the piles, a word or phrase that seems to sum up what the topics have in common. It may help to have other people look at your topics and try different groupings. Keep working until you have an organization you're satisfied with. The piles of topics with their labels can serve as a guide for the structure of your site.

make more sense than yours, or perhaps you can take elements from several organizational structures and combine them to create a more effective organization overall. At any rate, you shouldn't skip this step. Spending time on organization before you begin building the site can save costly redesigns later.

PROCESS FOR DESIGN PRELIMINARIES

As you prepare to create your Web site, keep these preliminary steps in mind:

1. If you haven't been assigned a topic, look for something that interests you, that will lend itself to several types of content, and that will be worth the time and effort you'll put into it.
2. List the goals for your site, perhaps considering the goals of the organization behind the site or determining what you want your site to accomplish.
3. List the primary and secondary audiences that are likely to use your site.
4. Identify sites with similar topics, and study them to find features that would be useful for your site or that should be avoided on your site.
5. List possible content you'll want to include on your site, and ask others who may be involved in the site to list content they'd like to see included.
6. Create a content inventory, listing all the content that you already have for inclusion in the site, as well as content to be created.
7. Try doing content chunking to organize your topics into groups and to label the groups based on what they have in common.

ASSIGNMENT: SITE SPECIFICATION PROPOSAL

Professional Web designers frequently create site specification proposals for large-scale sites, so that their clients can see (and approve) their preliminary plans and so that all members of the design team can have a common understanding of the site's requirements. For your first assignment, create a set of specifications for your Web site. Describe the topic you'll be using, the overall goals for the site, the audiences that might visit, their needs and the ways this site will meet them, and other similar Web sites that you might use for models. You should also include an outline of the site's probable content and preliminary organization.

CHAPTER SUMMARY

Advance Planning

❑ Successful Web sites require advance planning, beginning with a clear understanding of the purpose for which the site has been developed and the probable audiences that will be using the site.

❑ Designers should visit Web sites on similar topics in order to anticipate potential problems and possible solutions that have been devised on other sites.

Inventorying Content

❑ Designers should list possible topics to be covered on their site and invite others concerned with the site to submit lists of their own.

❑ Designers should make an inventory of existing content that will need to be converted to digital form and placed on Web templates.

Organizing Content

❑ Organization schemes—exact or ambiguous—are used to identify the kind of organizational structure to be used within a Web site.

❑ Organizational structures define the ways users can navigate through the site's information.

❑ Designers can devise organizational schemes and structures for their sites by analyzing the natural relationships that exist among the groups of information they've developed.

SOURCES

Fleming, J. 1998. *Web navigation: Designing the user experience.* Sebastopol, CA: O'Reilly.

Lynch, P. J., and S. Horton. 2001. *Web style guide: Basic design principles for creating Web sites,* 2nd ed. New Haven, CT: Yale University Press.

Rosenfeld, L., and P. Morville. 1998. *Information architecture for the World Wide Web.* Sebastopol, CA: O'Reilly.

HTML Basics

Chapter
2

WEB CODE

Now that you have some ideas about the content and organization for your site, we can approach the fundamental backbone of any Web site: the code. In this chapter we'll discuss HTML and XHTML. I'll give you an overview of the way HTML works, as well as some practice in basic HTML tags. In future chapters we'll go over these tags and their context in greater detail, but the information in this chapter, along with that in the next chapter on Cascading Style Sheets (CSS), should be enough to get you going.

LEARNING CODE

Coding is at the root of any consideration of Web design—without codes the pages don't work! Although beginning Web writers may believe that a Web authoring program, such as Adobe's Dreamweaver or Microsoft's FrontPage, can do away with the necessity of learning Web code, the reality is quite different. A big part of Web writing still involves Web coding—that is, using HTML and CSS, as well as other, newer codes, such as XHTML and even XML. Web authoring programs are wonderful tools, and they can certainly simplify the process of creating Web pages, but they aren't foolproof. After all, as someone wise once said, "Nothing is foolproof to a sufficiently talented fool."

In a sense, Web authoring programs work from the assumption that you're already familiar with HTML and CSS. They won't keep you from writing invalid code (although they will tell you that the code is invalid if you know how to ask). They won't tell you the most efficient way to do something; you can use as much or as little code as you like to accomplish your purpose. They won't keep you from making errors that will prevent your graphics from being loaded or will make your links inoperative. And they won't prevent you from writing outdated code that won't pass accessibility or validation tests. The quality of your code will depend on your knowing the difference between valid and invalid code.

In general, then, knowing how HTML and CSS work will help you use authoring programs more effectively. And using an authoring program will help you design and create pages more quickly. Learning code is still necessary, but it's not as difficult as many people believe.

HTML TAGS

HTML History

According to Tim Berners-Lee (2000), one of the creators of the Web and the head of the World Wide Web Consortium (the Web's governing body), HTML was intended to be "a simple hypertext language that would be able to provide basic hypertext navigation, menus, and simple documentation such as help files, the minutes of meetings, and mail messages—in short, 95 percent of daily life for most people" (40–41). Berners-Lee and his colleagues at CERN, the European scientific laboratory, envisioned a Web that would be accessible to anyone who wanted to use it and a tagging language that would be simple enough that a writer could create it using a plain ASCII text program. They also wanted a language that would use hypertext organization, linking pages together in a way that would allow users to choose their own paths through information, rather than moving in a linear format designed by the author. The result was hypertext markup language, or HTML.

As a result of Berners-Lee's determination to have a simple tagging language, basic HTML tags are easy to write. In fact, you can learn enough HTML in an hour to have a barebones Web page ready to post. However, since Berners-Lee and his colleagues created HTML in the early 1990s, the language has gone through four more generations and is now migrating to a new language: XHTML.

SIDEBAR: HTML TERMINOLOGY

Tag: An instruction to browsers that specifies the structure of Web content. For example, `<p></p>` is a tag that specifies that a section of text is to be treated as a paragraph. In earlier versions of HTML (prior to HTML 4.01), some tags also specified style; however, these tags have been superseded by CSS properties.

Attribute: A word or symbol placed within a tag that affects the appearance or behavior of an element on a page.

Block-level tag: An HTML element that begins on a separate text line and frequently has extra space inserted before it.

In-line tag: An HTML element that neither begins on a separate text line nor inserts extra space. It affects the appearance and behavior of an element within a line of text.

Text-only file: ASCII code without any of the formatting provided by a word processor. Text-only files are readable on most computers and don't require any special software (e.g., a word processing program) in order to be shown on the screen.

The Nature of HTML

HTML is a tagging language—a subsection of a much larger and more complex language, SGML (standard generalized markup language), which is used to make complicated technical documents readable in a variety of formats. HTML specifies the structure of your page—paragraphs, headings, lists, and so on—using a limited number of tags to identify your content. Since all Web sites use the same tags, all browsers can read the pages that use those tags, no matter what type of computer they were created with or what type of server they're mounted on. Again, according to Berners-Lee (2000), "A philosophical rule [that guided the creation of HTML] was that HTML should convey the structure of a hypertext document but not details of its presentation. This was the only way to get it to display reasonably on any of a very wide variety of different screens and sizes of paper" (41).

In general, this ideal has remained true: basic HTML tags are readable by all browsers, although they may not be rendered on the screen in exactly the same way. For example, if you place your text inside these tags—`<h1></h1>`—browsers will show a level-one heading, usually bolded and in a large font size. However, the font size itself may differ slightly from one browser to another and from one computer screen to another, depending on variables like the resolution of the screen or the default font size specified in the browser. Although your Web pages should be rendered similarly in all browsers, they may not look exactly the same, and trying to make them look identical everywhere is usually wasted effort. You want a design that will look more or less the same on the most recent versions of major browsers, but you probably can't make your pages look absolutely identical under all circumstances.

HTML BASIC INFORMATION

All HTML tags are written between less-than and greater-than signs (angle brackets), which enclose the letter *p* in this example:

```
<p>
```

Container Tags

Many HTML tags are **container tags**, that is, they have both an opening and a closing tag:

```
<h1> A Level-One Heading </h1>
```

The text ("A Level-One Heading") is *contained* between the two tags and is affected by them. Think of the opening tag as a switch that turns on the effect and the closing tag as a switch that turns off the effect.

The opening and closing tags use exactly the same word or symbol (e.g., h1), but the closing tag has a forwardslash before the word or symbol (e.g., /h1) in order to differentiate it from the opening tag. If you're using one of these

SIDEBAR: XHTML ALERT: TAGGING DIFFERENCES

XHTML is very similar to HTML in most ways, but there are some differences that you should be aware of. For example, although HTML is not case sensitive (i.e., you can write HTML in either capital or lowercase letters), all XHTML tags must be lowercase. In addition, XHTML requires that all opening tags have a closing tag. This requirement has meant that tags like `<br>` have had to be slightly reformatted. In XHTML the line break tag is written like this: `<br />`. The space before the slash isn't strictly necessary for the tag; however, it will prevent older browsers that don't understand XHTML from displaying the tag as if it were part of the text. Finally, XHTML requires all attribute values to be in quotation marks. We'll discuss XHTML in more detail later in this chapter.

container tags, **you must include the closing tag as well as the opening tag.** If you forget and leave off the closing tag, your text may not show up in the browser or may not have the style you expect. In HTML some of these closing tags are optional; in XHTML none of them are. Because all Web pages will eventually be written in XHTML, it's best to get into the habit of including close tags for all container tags (see *XHTML Alert: Tagging Differences*).

Not all HTML tags are container tags; some (like the line break tag, `<br>`) require only an opening tag. (It's still called an opening tag even if there's no closing tag, because it opens the HTML command.) If you don't remember whether the tag you're using needs a closing tag, check the tables provided at the end of each chapter. And if you're writing XHTML, remember to use the format required for these tags.

Attributes

Some tags also include *attributes,* which are words or symbols that affect the appearance or behavior of the content included between the tags. Here's an example:

```
<p class="red">
```

Here `<p>` is the opening tag, and `class` is an attribute that indicates what aspect of the paragraph you want to deal with. In this case you're applying a class, or set of styles, that you've defined on your style sheet. More than one attribute can be included in a tag; for example, you could have

```
<p class="red" style="font-size:medium">
```

However, all the attributes must come after the opening tag symbol (in this case p) and before the closing angle bracket and must be separated by spaces. Not all tags can include attributes; we'll cover the ones that can as we go through the HTML tags.

Many of the attributes that were in use in earlier versions of HTML were **deprecated** (i.e., made obsolete or invalid) in HTML 4.0 with the adoption of CSS. (There's a list of deprecated attributes in appendix 6.) There are two good reasons to avoid these deprecated attributes: the effect that the attribute was originally used to produce can now usually be produced more easily and efficiently using CSS, and the attribute itself will disappear from XHTML 2.0, which is not compatible with earlier versions of HTML.

Values

Attributes frequently have *values,* that is, they include further information like style or location. The value part of an attribute comes after the equals sign:

```
<p class="red">
```

Values should always be placed in quotation marks. Although these marks aren't required for all HTML values, they *are* required for all XHTML values, so once again, it's a good idea to get into the habit of using them.

Nesting Tags

Sometimes you'll have a series of tags around a particular paragraph or text section. For example, you might have

```
<p><a href="intern.html">Internship Program</a><p>
```

The main thing to remember here is that **the tag that opens first closes last.** In other words, if `<p>` is the first opening tag, then `</p>` should be the last closing tag. Your closing tags should repeat your opening tags in reverse order. Not all tags can nest like this, but several can.

WRITING HTML

As I indicated earlier, you can write your HTML in any kind of text program, from a full-featured word processor to the simple text-only programs that come with your computer's operating system (Notepad for Windows; TextEdit for Macintosh OS X). There is one big advantage to using these simple, text-only programs, however: **your HTML must be saved as a text-only (ASCII text) file.** If you forget and save your HTML as a Microsoft Word file or a WordPerfect file or any file other than text only, the browser will not be able to read your code. Since TextEdit and Notepad are both text-only programs, you can't slip up and forget to save the file as text only. Web authoring programs (such as Dreamweaver and FrontPage) and HTML editing programs (such as BBEdit or HomeSite) also produce text-only files. In fact, any file with the `.html` or `.htm` extension is text only (or should be).

Spacing HTML

Browsers pay no attention to spaces between your tags. Read through this first example:

```
<h1>Web Standards</h1> <p>Web standards weren't a big
concern during the Web's early days in the 1990s. Then
Web design was wide open to any use of HTML tags.
Designers came up with a variety of methods to achieve
things that HTML didn't allow them to do. For example,
since the default in HTML was single spacing, some
designers used a "design hack" in which they inserted
single-pixel gifs into their text in a way that forced
the lines to emulate double spacing.</p> <p>One of the
troubles with this and other design hacks was that
they could easily be messed up if you had a typo in
your code entry. Then you'd be stuck trying to track
down exactly what you'd done in the midst of a huge
stack of tags that were very hard to proofread.</p>
```

Code written like the first example will look exactly the same in the browser as code written like this second example:

```
<h1>Web Standards</h1>

<p>Web standards weren't a big concern during the
Web's early days in the 1990s. Then Web design was
wide open to any use of HTML tags. Designers came up
with a variety of methods to achieve things that HTML
didn't allow them to do. For example, since the
default in HTML was single spacing, some designers
used a "design hack" in which they inserted single-
pixel gifs into their text in a way that forced the
lines to emulate double spacing.</p>

<p>One of the troubles with this and other design
hacks was that they could easily be messed up if you
had a typo in your code entry. Then you'd be stuck
trying to track down exactly what you'd done in the
midst of a huge stack of tags that were very hard to
proofread.</p>
```

The big difference between these two examples, is that the second one is a lot easier to review quickly.

For example, suppose I left off the / on the `</h1>` close tag and then looked at the text in a browser. It would look like this:

Web Standards
Web standards weren't a big concern during the Web's early days in the 1990s. Then Web design was wide open to any use of HTML tags.

Designers came up with a variety of methods to achieve things that HTML didn't allow them to do. For example, since the default in HTML was single spacing, some designers used a "design hack" in which they inserted single-pixel gifs into their text in a way that forced the lines to emulate double spacing.

One of the troubles with this and other design hacks was that they could easily be messed up if you had a typo in your code entry. Then you'd be stuck trying to track down exactly what you'd done in the midst of a huge stack of tags that were very hard to proofread.

Everything on the page would come up in bold and a large size (i.e., the h1 heading style) because I didn't turn off the h1 tag. I'd want to debug that code to find out what had gone wrong, and it would be a lot harder to find the error in the first example, where all my lines run together, than in the second example, where each paragraph is set off so that I can look at it separately.

Block-Level Tags

If you want to put vertical space between elements on your Web page, you'll need to insert a block-level tag rather than just hitting the return key as you would on a word processor.

All **block-level tags** begin on a new line; sometimes extra vertical space is inserted as well (as with the paragraph and heading tags), but not always (as with the list item tags). Occasionally, beginning Web writers will insert line-break tags after list items or paragraphs to make them begin on a new line, but this is unnecessary since all block-level elements with start on a new line anyway. The following tags are block level and will always begin on a new line:

```
<div>        <h1>
<hr>         <h2>
<p>          <h3>
<li>         <h4>
<ol>         <h5>
<ul>         <h6>
<blockquote>
```

SIDEBAR: THE CASE OF BR

The line break tag `<br>` is technically an in-line element since it affects the behavior and appearance within a line of text. However, it does create a new line. You can use `<br>` when you want a new line but don't want any extra vertical space (e.g., if you are typing lines of poetry). Remember, in XHTML the tag is written as `<br />`.

SIDEBAR: THE WHITE SPACE "BUG"

Although using white space to separate your code and make it more readable is a good idea, it can, unfortunately, produce occasional problems in Internet Explorer. With this browser (through version 6.0), white space in your markup can sometimes produce slight gaps in your layout. These gaps may or may not become problems; they're most annoying in table layouts, where the table cells need to fit tightly against each other. If you use CSS layout, as I demonstrate in chapters 10 and 11, these gaps shouldn't be a significant problem. However, if you find that Internet Explorer is showing gaps that you didn't design in your pages, try removing the white space from your markup.

In-Line Tags

Unlike block-level tags, **in-line tags** do not begin on a separate line and do not insert vertical space. These tags affect the behavior and appearance of elements within a particular line of text. The following are in-line tags and do not begin a new line:

```
<a>
<img>
<span>
<object>
```

Writing HTML Tags

HTML isn't case sensitive, that is, browsers will recognize your tags whether they're in capital letters or lowercase. However, once again, this practice changes in XHTML; all XHTML tags must be in lowercase. Thus, it's a good idea to use lowercase for your tags whether you're writing HTML 4.01 or XHTML because you may be converting HTML pages to XHTML in the near future.

The best way to learn HTML is to start writing it, which is what we'll do in the following section. However, keep this in mind: HTML requires trial and error. Your first attempts may seem less than stellar, but once you get the hang of it, it won't seem so difficult. Take a problem-solving approach: How can I get this resource to do what I want, and how can I use this resource to make my message more effective?

CREATING AN HTML PAGE

Now that you have some basic information about HTML, let's try writing a simple HTML page. I won't go into great detail about the tags you'll be using; we'll talk more about them in future chapters. Moreover, the page you write won't look very fancy in a browser because we haven't covered CSS yet. But you can write a basic HTML page now and then set up the style sheet to make it more effective later on.

Head Section

Every HTML page is divided into two major sections: head and body.

The information in the **head section** is given to the browser and indicates how certain parts of the page are to be displayed. The information included in your head section includes the following tags, which we examine in order:

```
<!DOCTYPE HTML PUBLIC "-//W3C//DTD HTML 4.01
    Transitional//EN" "http://www.w3.org/TR/1999/
    REC-html401-19991224/loose.dtd">
<html>
<head>
    <title></title>
</head>
```

DOCTYPE. The Doctype tells the browser what version of HTML or XHTML is being used on this page, which will influence the way in which the page is rendered on the computer screen. Doctype is actually an XML tag and represents the first step in converting Web pages to the XML format (*extensible markup language* is discussed later in this chapter). Here we're using the Doctype for HTML 4.01 Transitional, the last version of HTML to be released by the World Wide Web Consortium. There are other Doctypes for HTML 4.01, including Strict, which requires closer adherence to the standards for this version of HTML, and Frameset for pages that use frames (but see the information about XHTML and frames later in this chapter). If you're writing your page using HTML, you'll most likely use the Doctype for either HTML 4.01 Transitional or HTML 4.01 Strict. The format here is the required format for the Doctype; it's the only tag that still uses capital letters in XHTML.

<html>. The html tag tells the browser that this page is written in html. This tag isn't absolutely required in HTML 4.01, but it's a good idea to include it. There's a slightly different tag used with XHTML, which is required. We'll cover that tag later in this chapter.

<head>. The information you provide in the head section doesn't appear in the browser window, but it includes information the browser needs in order to identify the page, such as the page's title. It can also include information about styles to be used on the page, which we'll discuss in the next chapter on CSS.

<title>. The text you place between the title tags will appear in the title bar at the top of the browser window. Don't confuse the title with a page heading—headings require a different format and must be placed in the body section of the page. In Figure 2–1 "HTML Basics" is the page title. Here are some things to keep in mind about titles:

❑ Titles should be short but specific. For example, "Page One" won't tell your reader the subject of the page, and it will be even more mystifying in a bookmark file. Use a word or two that will identify the subject of your page, such as "Communication Scholarships."
❑ You can't format the title in any way; each browser will provide its own formatting. So don't include any styles or images in the title.
❑ Don't use colons or slashes in your title since these symbols are used by some operating systems for other purposes.

</title></head></html>. Three of the tags used in the head section require close tags; the Doctype tag does not. The <html> close tag, </html>, comes after the close tag for the body section, </body>.

EXERCISE: CREATE YOUR HEAD SECTION

You can begin your page by creating your head section. Here we're going to use an XHTML head section, which is slightly different from the one just described. (We'll talk more about XHTML later in this chapter.) Using a text-only file, type in the following code:

```
<!DOCTYPE html PUBLIC "-//W3C//DTD XHTML 1.0
    Transitional//EN" "http://www.w3.org/TR/xhtml1/
    DTD/xhtml1-transitional.dtd">
<html xmlns="http://www.w3.org/1999/xhtml">
<head>
    <title></title>
</head>
```

Between <title> and </title>, type a title that describes the content of your page. The Doctype tag should be typed on a single line—it's been broken in two here by space constraints. Incidentally, once you have the Doctype typed in correctly, you can copy and paste it in future pages so that you don't have to keep typing in all that code. When you've finished typing in your head section, save your page, placing the .html extension at the end of the file name.

Body Section

After the head section comes the body section. The information in the **body section** will actually appear in the browser window. A typical body section might look like this:

```
<body>
<h1>HTML Tags</h1>
<p>
All HTML tags are written between less-than and
greater-than signs (angle brackets). Many HTML tags
are container tags, that is, they have both an
opening and a closing tag. All the content that
appears between these tags will be affected by them.
Think of the opening tag as a switch that turns on
the effect and the closing tag as a switch that turns
off the effect.
</p>
<p>
The opening and closing tags use exactly the same
word or symbol (e.g., h1), but the closing tag has
a forwardslash before the word or symbol (e.g., /h1)
in order to differentiate it from the opening tag.
If you're using one of these container tags, you
must include the closing tag as well as the opening
tag. If you forget and leave off the close tag, your
text may not show up in the browser, or it may not
have the style you expect it to have because you
didn't turn the effect off after you turned it on.
In HTML some of these close tags were optional; in
XHTML none of them are. Since all Web pages will
eventually be written in XHTML, it's best to get
into the habit of including close tags for all
container tags.
</p>
</body>
</html>
```

<body></body>. The body section of your page includes everything that visitors to your Web site will see in the main body of the page below the title bar: the text, images, and links to other pages. The bulk of your design will be done in the body section, and all of your page content should be placed after <body> but before </body>.

<h1></h1>. Headings in HTML are indicated by the h tags; they range from h1 (the highest-level heading) to h6 (the lowest-level heading). The text between <h1> and </h1> will show up in the default style that the browser uses for the highest-level heading because I haven't applied any styles to my headings yet.

EXERCISE: CREATE YOUR BODY SECTION

Now you'll add the content for your Web page in the body section. Begin by typing the following tags below your head section (although we're creating an XHTML page, the body section uses exactly the same tags as an HTML page):

```
<body>
</body>
</html>
```

After `<body>` but before `</body>`, type the following tags:

```
<h1></h1>
<p></p>
```

If you already have a heading in the content you're going to use for this page, copy and paste it after `<h1>` but before `</h1>`. If you don't have a heading to use, create one now and insert it between `<h1>` and `</h1>`.

If you have content that you can use from your content inventory, copy a paragraph of text and paste it on your Web page after `<p>` but before `</p>`. If you don't have a paragraph of text to use, create one now and insert it between `<p>` and `</p>`.

If you have another heading, enter `<h1>` before it and `</h1>` after it. If you have more than one paragraph, enter `<p>` before each one and end each with `</p>`.

Save your page again.

`<p></p>`. Paragraphs are indicated by the p tag; although the closing `</p>` tag isn't required in HTML, you should always include it since it is required in XHTML. The default styling (as you'll see in Figure 2–1) is to separate paragraphs with vertical space, similar to the usual style in business letters.

`</html>`. The closing `</html>` tag ends the page.

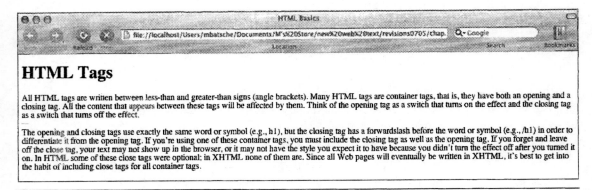

FIGURE 2–1   Page with text only

Adding an Image

Images can be placed in your body section using the `img` tag. There are several attributes that can be used with images and some fine points about finding and placing them that we'll go into in chapter 6, "Working with Images." However, I'll introduce the basic image tag here so that you can see how it works with the other basic HTML tags. It looks like this:

```
<img src="compwriter.gif">
```

img. The basic tag for adding a graphic to your Web page is `img`, which stands for *image*. It must be used with the `src` attribute. Notice that `img` is an empty tag, that is, it has no close tag. If you were writing this tag for an XHTML page, the format would be `<img src="compwriter.gif"/>`.

src. The `src` attribute stands for *source*. The value for the attribute is the file name for your graphic; in this case the file name is `compwriter.gif`. The file name must be placed in quotation marks.

If I add a graphic to the basic page body just created, the page will look like this:

```
<body>
<h1>HTML Tags</h1>
<p>
All HTML tags are written between less-than and
greater-than signs (angle brackets). Many HTML tags
are container tags, that is, they have both an
opening and a closing tag. All the content that
appears between these tags will be affected by them.
Think of the opening tag as a switch that turns on
the effect and the closing tag as a switch that turns
off the effect.
</p>
<img src="compwriter.gif">
<p>
The opening and closing tags use exactly the same
word or symbol (e.g., h1), but the closing tag has
a forwardslash before the word or symbol (e.g., /h1)
in order to differentiate it from the opening tag.
If you're using one of these container tags, you
must include the closing tag as well as the opening
tag. If you forget and leave off the close tag, your
text may not show up in the browser, or it may not
have the style you expect it to have because you
didn't turn the effect off after you turned it on.
In HTML some of these close tags were optional; in
XHTML none of them are. Since all Web pages will
eventually be written in XHTML, it's best to get
```

```
into the habit of including close tags for all
container tags.
</p>
</body>
```

In a browser the page with the image would look like Figure 2–2.

EXERCISE: ADDING AN IMAGE

If you have a graphic in your content inventory, you can add it to your page now. Note that graphics for Web pages can use only three file formats: GIF, JPEG, or PNG. If the graphic file you want to use isn't in one of these three formats, you'll need to take it into your graphics program and save it in one of these three Web file formats. (I'll cover these file formats in more detail in chapter 6.). To add your graphic to your page, make sure it's in the same file folder as your Web page on your computer. Then type the following tag in the location where you'd like the graphic to appear:

```
<img src="yourgraphicfilename">
```

Instead of `yourgraphicfilename`, type the name of your graphic file. Be sure to add the file extension (i.e., `.gif`, `.jpg`, or `.png`).

Adding a Link

Links to other pages are a fundamental part of all Web sites. They use different formats depending on whether the link is to another page on the same Web site, to a page on a different Web site, or to another location on the same Web page.

HTML Tags

All HTML tags are written between less-than and greater-than signs (angle brackets). Many HTML tags are container tags, that is, they have both an opening and a closing tag. All the content that appears between these tags will be affected by them. Think of the opening tag as a switch that turns on the effect and the closing tag as a switch that turns off the effect.

The opening and closing tags use exactly the same word or symbol (e.g., h1), but the closing tag has a forwardslash before the word or symbol (e.g., /h1) in order to differentiate it from the opening tag. If you're using one of these container tags, you must include the closing tag as well as the opening tag. If you forget and leave off the close tag, your text may not show up in the browser, or it may not have the style you expect it to have because you didn't turn the effect off after you turned it on. In HTML some of these close tags were optional; in XHTML none of them are. Since all Web pages will eventually be written in XHTML, it's best to get into the habit of including close tags for all container tags

---

FIGURE 2–2   Page with text and image

All of these formats will be covered in chapter 8, "Linking Pages." However, here I'll show you the basic link tag:

```
<a href="http://www.nypl.org/styleguide/"></a>
```

<a></a>. The basic link tag is a, which stands for *anchor,* and must include the attribute href. Between <a> and </a>, you'll place explanatory information about the link, such as the name of the Web page you're linking to, like this:

```
<a href="http://www.nypl.org/styleguide/">The New
York Public Library's XHTML Style Guide</a>.
```

href. The href attribute is required with a; it stands for *hypertext reference.* The value for href is the URL of the page to which you're linking, that is, its Web address, which can be found in the address field at the top of your browser window. This value should always be placed in quotation marks.

If I add a link to the body of the page I created earlier, it will look like this:

```
<body>
<h1>HTML Tags</h1>
<p>
All HTML tags are written between less-than and
greater-than signs (angle brackets). Many HTML tags are
container tags, that is, they have both an opening and
a closing tag. All the content that appears between
these tags will be affected by them. Think of the
opening tag as a switch that turns on the effect and
the closing tag as a switch that turns off the effect.
</p>
<img src="compwriter.gif" />
<p>
The opening and closing tags use exactly the same
word or symbol (e.g., h1), but the closing tag has a
forwardslash before the word or symbol (e.g., /h1) in
order to differentiate it from the opening tag. If
you're using one of these container tags, you must
include the closing tag as well as the opening tag.
If you forget and leave off the close tag, your text
may not show up in the browser, or it may not have
the style you expect it to have because you didn't
turn the effect off after you turned it on. In HTML
some of these close tags were optional; in XHTML none
of them are. Since all Web pages will eventually be
written in XHTML, it's best to get into the habit of
including close tags for all container tags.
</p>
<a href="http://www.nypl.org/styleguide/">The New
York Public Library's XHTML Style Guide</a>
```

```
</body>
</html>
```

EXERCISE: ADDING A LINK

Now you can add a link to your basic Web page. For now, you'll create a link to another Web site since you probably don't have any other pages on this particular site to link to yet. Find a Web site that includes information related to the subject of your page by going to a search engine (e.g., Google or Yahoo) and typing in keywords related to the page topic. When you find a page that seems to include appropriate information, copy the URL that appears in the address field at the top of your browser window. Then type the following tags on your page where you want the link to occur:

```
<a href=></a>
```

After the `href=` paste the URL that you copied from the browser window. Then type quotation marks before and after it, so that it looks like this (instead of URL, you'll have the complete address of the Web page that you copied from your browser):

```
<a href="URL"></a>
```

Now, after `<a href="URL">` but before `</a>`, type the name of the Web site that the URL belongs to. The result should look something like this (with your own URL and the name of the Web site substituted for the example I'm using here):

```
<a href="http://www.nypl.org/styleguide/">The
    New York Public Library's XHTML Style Guide</a>
```

We now have a complete Web page, with text, image, and a link. The code for the entire page looks like this:

```
<!DOCTYPE HTML PUBLIC "-//W3C//DTD HTML 4.01
    Transitional//EN" "http://www.w3.org/TR/1999/
    REC-html401-19991224/loose.dtd">
<html>
<head>
    <title>HTML Basics</title>
</head>
<body>
<h1>HTML Tags</h1>
<p>
All HTML tags are written between less-than and
greater-than signs (angle brackets). Many HTML tags are
container tags, that is, they have both an opening and
a closing tag. All the content that appears between
these tags will be affected by them. Think of the
```

```
opening tag as a switch that turns on the effect and
the closing tag as a switch that turns off the effect.
</p>
<img src="compwriter.gif" />
<p>
The opening and closing tags use exactly the same
word or symbol (e.g., h1), but the closing tag has a
forwardslash before the word or symbol (e.g., /h1) in
order to differentiate it from the opening tag. If
you're using one of these container tags, you must
include the closing tag as well as the opening tag.
If you forget and leave off the close tag, your text
may not show up in the browser, or it may not have
the style you expect it to have because you didn't
turn the effect off after you turned it on. In HTML
some of these close tags were optional; in XHTML none
of them are. Since all Web pages will eventually be
written in XHTML, it's best to get into the habit of
including close tags for all container tags.
</p>
<a href="http://www.nypl.org/styleguide/"> The New
     York Public Library's XHTML Style Guide</a>
</body>
</html>
```

In a browser the page would look like Figure 2–3.

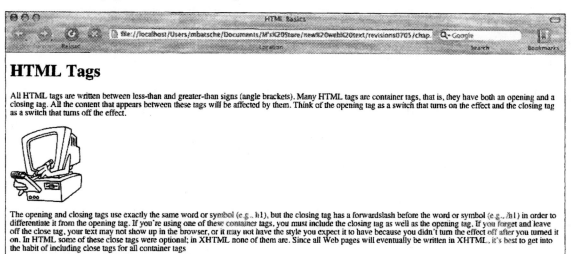

FIGURE 2–3 A basic page

CHECKING WITH A BROWSER

As you look at the figures in this and other chapters, you'll see that there are some slight differences among them. That's because I used different browsers to look at them (Safari, Camino, and Firefox). You, too, should check your pages frequently to see how they're showing up in standard browser windows. And don't limit yourself to a single browser, even if that's all you use when you surf the Web. Other people will be looking at your pages in different Web browsers, and you need to know what they'll be seeing.

Take this opportunity to look at any page you wrote using the tags described so far. Remember to save your pages with the .html extension. You're not going online now because your page isn't mounted yet; instead, you'll look at it off-line, which will give you some idea of how it would look on the Web.

SIDEBAR: NAMING AND SAVING FILES

You must save your HTML or XHTML file before you can look at it in a browser, but there are some conventions in file naming that you need to be aware of. First of all, to repeat, your HTML or XHTML files need to be saved in text-only (ASCII text) format in order to be readable on the Web. You must also use the .html extension after the name of your HTML files.

EXTENSIONS

Extensions are a series of letters at the end of a file name, usually separated from the file name by a period—for example, scholarships.html. The extension tells the browser what kind of information is stored in the file. Thus, .html indicates that the file is written in HTML, whereas .gif would indicate that the file is a GIF graphic. Although it's sometimes assumed that PC users save as .htm and Mac users save as .html, the distinction actually depends on the version of Windows that PC users are using. Windows 3.1 or earlier can't use a four-letter extension like .html—it must be .htm. But a later version of Windows or a Mac can use .html. Since few people are using the older version of Windows at this point, .html is the standard. Incidentally, XHTML files are also saved with the .html file extension because many browsers can't handle files that end in .xhtml.

NAMING CONVENTIONS

When you name your files (including HTML, graphics, sound—any files), there are two things to remember:

1. Keep the file names short, usually no more than twelve characters. Some browsers can't read longer file names.
2. **Leave no spaces between the words in your file names.** A name such as com file.html wouldn't work as a file name; you'd have to make it comfile.html. You need to be careful about this because some browsers won't read files that have spaces in the names.

EXERCISE: SEEING YOUR PAGE IN A BROWSER

After you've created and saved your basic HTML page, you'll want to look at it in a Web browser so that you can get a preview of the way it will look after it's been mounted on the Web. Use the following steps to check your page in your Web browser:

1. If you haven't already done so, save your Web page.
2. Open your Web browser.
3. Open your Web page in the browser: go to the File menu and select the command Open File. Your browser will probably show you a window containing your computer's desktop or the contents of your hard disk.
4. Locate your Web page file and click on the name to open it in the browser.
5. Check your page to make sure all of your content appears. If it doesn't, you'll need to look at your HTML code to make sure you don't have any typos.

XHTML

With the introduction of XML (extensible markup language), HTML has undergone another set of changes, becoming a new language—XHTML. HTML has served the Web well, but its capabilities are narrow. It cannot convey some nontextual content, such as mathematical formulas, and its applicability for database-driven Web sites has been very limited. Using XML, organizations can not only use databases more effectively, but they can also develop their own markup languages (we'll discuss XML in more detail later in this book). XHTML provides a bridge between the older, HTML-based Web and the new, XML-based version.

Fortunately, XHTML is similar to HTML in many ways, but there are some significant differences that we'll summarize here. First, just like the HTML Doctype, the XHTML 1.0 Doctype comes in three different versions: strict, transitional, and frames (which will disappear in future versions of XHTML). But the format is somewhat different. Because it is an XML markup language, XHTML documents should begin with an XML version declaration. However, several current browsers cannot handle the XML declaration, so we begin with the Doctype, which looks similar to the one used for HTML:

```
<!DOCTYPE html PUBLIC "-//W3C//DTD XHTML 1.0
    Transitional//EN" "http://www.w3.org/TR/xhtml1/
    DTD/xhtml1-transitional.dtd">
```

Because XHTML is an XML language, however, it requires one more line of code at the top: an XML namespace. The namespace in XML defines the elements and attribute names that are part of the particular language being used, in this case XHTML 1.0 transitional. It looks like this:

```
<html xmlns="http://www.w3.org/1999/xhtml">
```

This tag takes the place of the old `<html>` tag; it identifies where the definitions for the tags and attributes used in XHTML can be found. With the two parts together, the XHTML Doctype definition looks like this:

```
<!DOCTYPE html PUBLIC "-//W3C//DTD XHTML 1.0
    Transitional//EN" "http://www.w3.org/TR/xhtml1/
    DTD/xhtml1-transitional.dtd">
<html xmlns="http://www.w3.org/1999/xhtml">
```

What comes after this Doctype? The head section, followed by the body section. After the Doctype, the page structure is exactly the same as it is in HTML. But there are some other differences between HTML and XHTML to be aware of.

❑ First of all, XHTML is much more demanding than HTML was. Misused or missing tags are sometimes overlooked by HTML browsers, but they aren't in XHTML browsers. For example, elements must be nested correctly in XHTML (i.e., the first element opened is the last element closed). Although this was also true in HTML, incorrect nesting with XHTML can result in the page not being visible in a browser.

❑ All XHTML tags require close tags. This presents no problem with container tags, of course, but empty elements (i.e., elements without close tags) are another matter. The way to handle this is to put a forwardslash at the end of the empty element tag. For example, the `<img src="picture.jpg">` discussed earlier becomes `<img src="picture.jpg"/>`. The space before the slash isn't required, but using it will fool older HTML browsers so that they render the tag the way you want it to be rendered.

❑ XHTML is case sensitive; all tags and attributes must be written in lowercase.

❑ All attribute values must be placed in quotation marks in XHTML.

❑ Frames are not part of the XHTML 2.0 specification, which is a good reason not to use them.

There are some other rules for XHTML that you may or may not encounter as a beginning designer. For example, all attributes in XHTML require values, but few attributes remain that don't have values because values like `nowrap` and `noshade` have been deprecated. You're most likely to run into these attributes with forms; the solution is simply to repeat the attribute name, for example, `checked="checked"`. In addition, if you use either & or < as characters on your page (i.e., not as angle brackets around your tags), you must encode them using ASCII characters:

```
&amp;
<&lt;
```

For more information on these ASCII character references, see appendix 5. However, unless you're writing mathematical copy, you may not run into this situation.

HTML or XHTML?

Since HTML and XHTML are so similar, you may wonder which one you should be using. In general, you should become familiar with both languages. Because existing HTML sites will continue to be written in HTML for the foreseeable future, all Web designers need to have a working knowledge of that language. However, when you create new sites, it's best to create them using XHTML in order to take advantage of the next generation of browsers. As Chuck Musciano and Bill Kennedy (2002) suggest, "Given the resources and opportunity, you should try to create XHTML-compliant pages wherever possible for the sites you are creating. Certainly you should choose authoring tools that support XHTML and give you the option of generating XHTML-compliant pages" (514).

OTHER HTML TAGS

Browser-Specific Tags

During the 1990's both Netscape and Microsoft Internet Explorer introduced tags that could be read only by those particular browsers, that is, they were browser-specific tags. For example, Explorer introduced a marquee tag that created a scrolling display but was readable only by the Explorer browser. In each case the browser manufacturers hoped that more designers would use their particular tags so that more users would use their browsers. However, using tags that could be recognized only by one particular browser seemed to defeat the whole purpose of a universal tagging language, and many Web designers refused to use them. These tags are still in existence; in fact, some of them have been used in newer versions of HTML and XHTML, making them no longer browser specific. In this text I do not include tags that remain browser specific, beyond listing them in appendix 6 so that you can avoid them. Using browser-specific tags automatically limits the number of readers who can use your pages.

Deprecated Tags

With the adoption of CSS as part of HTML 4.01 and XHTML 1.0, many older HTML tags and attributes have been deprecated, that is, they have been "outdated by newer constructs" according to the World Wide Web Consortium. Beginning Web designers need to be aware of these deprecated tags. Even though they are still readable by most browsers, there are good reasons for avoiding them:

❑ There are easier and more efficient ways to accomplish your aims using CSS.

❑ In some cases deprecated tags may present accessibility problems for Web-enabled devices.

❑ Although these tags are readable in current Web browsers, there is no guarantee that they will remain so in the future. XHTML 2.0 is not backward compatible, meaning that it does not include deprecated tags that have been dropped from the XHTML specifications.

I've listed the deprecated tags and attributes in appendix 6 so that you can use the newer, more efficient code that replaces them.

PROCESS FOR CREATING A BASIC PAGE

Use the following steps to create a basic XHTML page:

1. Type the tags of the head section:

```
<!DOCTYPE html PUBLIC "-//W3C//DTD XHTML 1.0
    Transitional//EN" "http://www.w3.org/TR/xhtml1/
    DTD/xhtml1-transitional.dtd">
<html xmlns="http://www.w3.org/1999/xhtml">
<head>
    <title></title>
</head>
```

2. After `<title>` but before `</title>`, type your page title.
3. After `</head>` type the tags for your body section:

```
<body>
</body>
</html>
```

4. After `<body>` but before `</body>`,

❑ Type `<h1>`

❑ Type or paste your first heading

❑ Type `</h1>`

5. After `</h1>` (perhaps on a separate line),

❑ Type `<p>`

❑ Type or paste your paragraph text

❑ Type `</p>`

6. If you want to insert a graphic, type the following tag:

```
<img src="graphicfilename"/>
```

Insert the name of your graphic file (with its extension) in place of `graphicfilename`.

7. If you want to insert a link to another page, type the following tag:

```
<a href="URL"> Description of Web page</a>
```

Insert the actual URL of the chosen Web page in place of URL and the description of the page between `<a>` and `</a>`.

8. Save your file with the `.html` extension.

Note: Repeat steps 3 and 4 for any other headings and paragraphs. If you want subheadings, use h2 or h3 rather than h1.

ASSIGNMENT: WEB SITE

You can use the directions here to begin creating pages for your Web site because you can always make changes later using CSS. If you have pages in your content inventory that are largely text, you can begin setting them up now, writing or pasting headings, subheadings, and paragraphs into your code. We'll add some style rules in the next chapter, and you can continue to refine your design as you work on your pages.

CHAPTER SUMMARY

HTML Basics

❑ HTML conveys the basic structure of a Web page to make it readable in a variety of formats and across all platforms.

❑ All HTML tags are written between angle brackets. Some HTML tags are container tags, which require both an opening and a closing tag. If the closing tag is omitted, the effect of the tag may not be applied or may be applied to other elements.

❑ Attributes are words or symbols that are placed inside HTML tags and that affect the appearance or behavior of the information contained within those tags.

❑ Values define aspects of attributes, such as styles or locations. They are placed after an equals sign and should be enclosed in quotation marks.

❑ HTML tags may be nested inside one another in some instances. In these cases the tag that opens first is closed last.

❑ HTML must be written in ASCII text files (text-only files). These files can be produced by the text writers included with most computer operating systems, or they can be produced by HTML editors and Web authoring programs.

❑ Internal spacing of HTML doesn't affect its appearance in a browser. However, Web site designers may use spacing to make their code easier to check.

HTML Tags

❑ Block-level tags begin on a new line and sometimes have extra space inserted before them (e.g., `<p>`, `<h1>`).

❑ In-line tags do not begin on a separate line, nor do they insert extra space. They affect the appearance or behavior of elements within a line of text (e.g., `<a>`, `<img>`).

❑ All HTML pages include the same basic structure: a head section that conveys information to the browser that will affect the display of the page and a body section that includes the page content.

XHTML

❑ XHTML is a version of HTML that follows the conventions of XML. The page structure of XHTML differs from HTML in the head section but not in the body section.

❑ XHTML demands strictly valid code; all XHTML tags require close tags; XHTML is case sensitive, requiring all tags to be written in lowercase; all XHTML values must be placed in quotation marks; and XHTML has dropped frames from the specification.

Deprecated and Browser-Specific Tags

❑ Both Netscape and Microsoft have created browser-specific tags that can be understood only by the browsers for which they were created.

❑ HTML 4.01 and XHTML 1.0 deprecate several tags used in earlier versions of HTML. Although these tags may still be read by current browsers, newer versions of browsers may no longer recognize them.

SOURCES

Berners-Lee, T. 2000. *Weaving the Web*. New York: HarperBusiness.

Musciano, C., and B. Kennedy. 2002. *HTML and XHTML: The definitive guide*, 5th ed. Sebastopol, CA: O'Reilly.

Cascading Style Sheet Basics

Chapter 3

CASCADING STYLE SHEETS

The emergence of Cascading Style Sheets (CSS), over the past five years has brought about a major change in the way Web pages are designed. In this chapter we'll cover the basic principles of CSS, including the format for creating internal and external style sheets and the required format for the style rules that go on those sheets. We'll also return to the plain vanilla HTML page created in the previous chapter and show you how to apply some simple styles to make it more appealing. But first, we need some background for CSS.

Web Standards

Writing code that was compliant with what are now called Web standards was not a serious issue in the early days of HTML. Browsers were designed to be forgiving and frequently overlooked errors that designers made in writing HTML. However, the introduction of Cascading Style Sheets and, more importantly, XHTML have changed attitudes regarding Web standards.

Prior to the introduction of Cascading Style Sheets in 1996, Web designers were limited in the kinds of design they could do using HTML, which was never intended to be used for page layout. Originally, Berners-Lee and the other early Web designers had assumed that the Web would be used largely by scientists who wanted to share data. Thus, HTML tags were created on the assumption that Web page designers would be posting large sections of text, with occasional illustrative diagrams and photographs. Elements like multiple columns, colors, different type fonts, and decorative graphics were simply not considered.

However, as more and more graphic designers began designing Web sites, there was a greater demand for more sophisticated design tools. Designers began devising—and sharing—various "design hacks," which used HTML tags for purposes other than the ones they were originally designed to fulfill. For example, since HTML didn't include any kind of layout positioning or spacing (not even margins), designers began using invisible tables to position their page elements.

A two-column table could create a left margin with a colored background that would be ideal for a menu, and a multicolumn table (with the borders turned off) could be used to create a grid similar to the grids used in print page design. Similarly, because it was difficult to control type fonts, designers created elaborate headings using decorative type in a graphics program like Photoshop, saved the headings as GIF images, and then placed them on their layout tables. These design hacks were widely embraced but did some damage to the way HTML fulfilled its original purpose of conveying the structure of information. And they were disastrous for accessibility—trying to read design tables with assistive technology is difficult at best.

Another problem in using HTML came about because graphic designers and site owners wanted their sites to look exactly the same in all browsers and all browser versions. In practice, this meant creating several different versions of sites, using different tags to accommodate each browser version. Site visitors encountered an opening page that began with an invisible "browser sniffer"—a JavaScript section that checked the version of the browser being used. Then visitors were sent to the version of the site that used the code that worked best for their browser. Obviously, keeping up all of these various site versions was time-consuming and expensive. Designers and Webmasters spent a great deal of time trying to make their sites perform identically in both the earliest browser versions and the most recent ones, and the multiple versions of each site consumed budgets and bandwidth. Many Web professionals complained about the waste of resources occasioned by the differing versions of HTML implementation in various browser versions.

The Web Standards Movement

CSS1 and CSS2 took care of many of the designers' complaints about the lack of page layout options. Margins and spacing became easy to set; different type fonts were now more available (although still dependent on the fonts loaded on a user's computer); and CSS positioning could accomplish most of what designers had once accomplished using layout tables. With CSS and HTML, designers could separate structure and format; create cleaner, more efficient code; and reach the ideal of an accessible Web. However, they could do these things only if browsers were developed that were compliant with the specifications of the World Wide Web Consortium (W3C) for HTML and CSS, and it took a few years for these browser versions to appear.

The first largely compliant browser versions were Netscape 6.0 and Explorer 5.0. Until those browsers were released, many designers continued to use a mixture of designs that were compliant and noncompliant with W3C standards, using tables for layout while using CSS properties for fonts and spacing, for example. And although using external CSS meant that sites could be designed easily for various browsers (only the style sheet needed to be changed rather than the HTML code), many designers continued to keep up separate sites for earlier browser versions even though many complained about the necessity for doing so.

At this time a group of designers and Webmasters formed the Web Standards Project (WSP) and began working with the major browser producers, trying to devise browsers that would render CSS and HTML according to the specifications established by the W3C. Once those more-compliant browsers became available, the WSP argued that all sites should use standards-compliant code and that designers should no longer produce noncompliant versions of sites for older browsers. The WSP pointed out that since the HTML code used on standards-compliant sites would be readable by all browser versions and only the style sheets would be missed by older browsers, using CSS would make it possible to produce sites that presented the same information for all users, although some users would see more attractive page designs than others. In other words, the WSP argued against the earlier ideal of identical site presentation in all browsers and browser versions. The WSP even supplied a code segment for designers to insert into their pages so that users of noncompliant browsers could be told that the appearance of the site would be much improved if they used a standards-compliant browser. The WSP contended that doing this would support manufacturers who created standards-compliant browsers and users who upgraded to the newer, more compliant versions, rather than rewarding those users and manufacturers who refused to do so.

To some extent this call for standards compliance has been gradually embraced by the Web design community. Most professional sites now use CSS rather than the older deprecated tags. However, some resistance to the Web standards movement still exists, particularly among designers who argue that table-based layout provides more options than CSS positioning. In some cases older sites involving several thousand pages would have to be completely redone to avoid tables and noncompliant code. For these sites it may make sense to introduce a new design for current information while maintaining the older design for pages that are archived or to use hybrid designs that keep simple tables for some layout while embracing CSS for other aspects of presentation.

Although complaints about converting older designs may arouse some sympathy, there's really no reason to *begin* a new site without using standards-compliant code. As more and more of the CSS specifications are incorporated into newer browser versions, there's less and less excuse for not using them. For the remainder of this book, we'll be working with standards-compliant design as far as possible, although as we'll see in chapter 12 ("Using Multimedia"), sometimes noncompliant code is necessary because of browser deficiencies.

SIDEBAR: MORE ON WEB STANDARDS

Several Web sites have information about current Web standards, but perhaps the best place to start is the World Wide Web Consortium's Web site at http://www.w3.org/. The Web Standards Project at http://www.webstandards.org/ can also provide current information as well as links to other sites on Web standards.

USING CASCADING STYLE SHEETS

CSS was added to HTML in 1996 and was designed to provide the features that writers and designers had found lacking in the original versions of HTML. CSS allows you to choose fonts, specify font sizes using familiar measurements rather than the somewhat arcane number system used in HTML, establish margins and line spacing, and, best of all, position elements on a page by specifying an exact location.

Style sheets also separate the presentation of pages (i.e., their appearance) from the structure of pages, as Berners-Lee and his colleagues had originally specified. The HTML tags specify page structure, that is, paragraphs, headings, and so on; CSS specifies presentation, that is, fonts, colors, spacing, and such. This separation is useful both for accessibility and for creating pages that are easily adaptable to various Web-enabled devices. The separation also greatly simplifies the code, making pages load more quickly in a browser; and style sheets make it easier to make changes quickly, thus simplifying site maintenance.

There are two CSS releases, CSS1 and CSS2 (CSS3 is currently under construction at the W3C). At the moment CSS1 has been implemented in most recent browser versions, and many features of CSS2 have been implemented, but not all of them. As XHTML-enabled browsers become the standard, more and more of the CSS properties will be available to Web site designers.

Perhaps the easiest way to think of CSS is to compare it to the style sheets in word processing programs. In most full-featured word processors and page layout programs (such as Microsoft Word, WordPerfect, QuarkXpress, and Adobe InDesign), you can specify a certain combination of style features for each part of your document. For example, if you want your top-level headings to be twelve-point bold Helvetica, you can define that heading style in your style sheet. Then every time you want to create a new heading, you just click on the heading style and the text is automatically changed to twelve-point bold Helvetica. If you decide later that you'd like those headings to be in fourteen-point Helvetica rather than twelve, you can make the change in the style sheet and automatically change the style on all the top-level headings in your document.

Clearly, word processor style sheets are a great time-saver, but they also help to unify documents, ensuring that a particular element always looks the same. That can help your readers make sense of your information more quickly and also help them understand the organization of your page at a glance.

CSS gives you the same kind of ability to define your styles and keep them uniform throughout your Web site. Remember, you want your readers to be able to scan your pages, identifying information quickly. Consistent presentation helps them do that. Style sheets also allow you to change the default HTML styles established by browsers. Rather than the bolded oversize style that most browsers set for h1, for example, you can choose a different font, size, treatment, and even color. In addition, through classes and ids, CSS allows you to create elements that function like your own HTML tags.

Internal and External Style Sheets

You can create style sheets in several different ways using CSS. **External style sheets** are created on a separate page of your Web site; you write all of your style rules on that page, then save it as a text-only file with the extension `.css`, and link it to the pages where it's applied by inserting the following line of code in your head section: `<link rel="style sheet" type="text/css" href="filename.css">`. In this tag `rel` stands for *relationship*, defining what kind of document you're linking to your page; `type` is a MIME type (we'll discuss MIME types in greater depth in chapter 12); and `href` again stands for *hypertext reference* and gives the file name of your style sheet. The external style sheet will be applied to all the pages that have that `link` tag in the head section, as long as the `href` attribute has the file name of your style sheet as the value.

    Internal style sheets are placed in the head section of your page using these tags: `<style type="text/css"> </style>`. Your style rules are placed between the opening and closing tags and will apply only to the page on which they're placed. You can also use the `style` attribute with an HTML tag to apply a style to only one particular element on a page, e.g., `<p style="font-size: x-large;">`.

    These styles can override each other; that is, if you defined a paragraph style on your external style sheet and then redefined some properties of that style on your internal style sheet, the internal styles would override the external ones. And if you used a `style` attribute in a `<p>` tag to redefine some properties for one particular paragraph, these attribute styles would override both the external and internal styles. That's the cascade part of Cascading Style Sheets.

Linking Style Sheets with `@import`

In addition to the `<link>` tag just described, you can import external style sheets into the style section of your page. The code looks like this for an external style sheet named "mystyles.css":

```
<style type="text/css">
@import url("mystyles.css");
</style>
```

Although either method of linking to an external style sheet will have the same effect, this `@import` method works only in version 5.0 and above in Internet Explorer and version 6.0 Netscape. More recent browsers, such as Safari and Firefox, will read `@import` without difficulty.

    Since earlier browsers don't recognize the `@import` method, designers can use different style sheets for different browser types. Older browsers will recognize a style sheet linked with the `<link>` tag, while ignoring the style sheet linked with the `@import` rule. The style sheet referred to in the `<link>` tag can include basic CSS elements like fonts, text spacing, and colors that can be used in older browsers. The second style sheet linked by the `@import` rule can be used by newer, more-compliant browsers and can include more CSS rules. As Jeffrey Zeldman (2003) points out, "By taking this layered approach, you can support all users without blocking your content from anyone—and without the agony of browser detection" (227). The code for this two-style-sheet method looks like this:

```
<!DOCTYPE html PUBLIC "-//W3C//DTD XHTML 1.0
    Transitional//EN" "http://www.w3.org/TR/xhtml1/
    DTD/xhtml1-transitional.dtd">
<html xmlns="http://www.w3.org/1999/xhtml">
<head>
<title>Page Title</title>
<link rel="style sheet" type="text/css"
    href="oldstyles.css" />
<style type="text/css">
@import url(newstyles.css);
</style>
</head>
```

Older browsers will use the `oldstyles.css` style sheet in the `<link>` tag; newer browsers will use the `newstyles.css` style sheet in the `@import` rule because it comes later in the head section.

CSS BASICS

To understand what goes into the styles themselves, you will need to become familiar with the terms in the sidebar on CSS terminology. The format for CSS rules is the same wherever you place them:

SIDEBAR: CASCADING STYLE SHEET TERMINOLOGY

CSS rule: A line of CSS code that defines how an HTML element will look or behave in a browser (e.g., `p{color:red;}`). All rules have at least two parts: a selector and a property.

Style attribute: An attribute that applies a CSS rule to a particular HTML tag.

Selector: A word or set of characters in a CSS rule that identifies the elements to which a particular style will be applied. It can be an HTML tag (e.g., `p`) or a class or id selector.

Parent and child: Two kinds of selectors. Parent selectors enclose child selectors; for example, in `<h1><em>Important Heading</em></h1>`, `em` is the child of `h1`.

Property: The style aspect that's being defined. It includes the way an element looks (e.g., color, size) and the way it acts (e.g., position).

Value: The definition of the property (e.g., green, x-large).

Class: A special set of styles that can be applied to one or more selectors on a page.

Id: A special set of styles that can be applied to only one selector on a page.

```
selector{property:value; property:value;}
```

Your style rules can include as many properties as you want to apply to a particular element, but the properties and values must be separated by semicolons. In fact, it's a good idea to end your style rule with a semicolon in case you want to come back and add more properties later.

Notice that this format is quite different from the format for HTML attributes—there are no equals signs or quotation marks. In fact, if you write style rules with equals signs or quotation marks, they won't work! Obviously, not all values will work for all properties; for example, `font-size` could never be defined as `brown`. If you're ever confused about whether a property and a value will work together or whether a particular measurement (e.g., inches) will work with a particular property, you can consult the CSS property table at the end of most chapters.

CSS is based on the structure of documents, which means it takes advantage of the ***parent-child relationship*** of the elements in those documents. Parent elements enclose child elements; for example, in the following code, `h1` is a child of `div` because it is enclosed within it (`div` stands for *division*, a block-level element that we'll discuss more thoroughly in chapter 10).

```
<div>
<h1>A Level One Heading</h1>
</div>
```

The importance of these parent-child relationships will become more obvious as we look at other types of selectors.

Classes and Ids

Sometimes you may want to create special styles that you'll apply to some elements on a page but not to others. Say you wanted to emphasize some paragraphs but not all of them. You wouldn't want to use `p` as the selector when you defined the style because that would apply the style to all of the paragraphs on the page. Instead, you can create a special type of selector called a ***class***, which you can apply to the paragraphs you want to emphasize.

You define a class in a CSS style rule. The only difference in the format of the rule comes with the selector: the class name must begin with a period. The format for a class defnition is this:

```
.classname{property:value; property:value;}
```

A sample class rule might look like this:

```
.cool{color:blue;}
```

You can use classes with any HTML tag, as long as the properties in the class work with that tag. To use a class with an HTML tag, you insert the `class` attribute in the tag, like this: `<p class="cool">`.

You can apply the same class to several different tags on the same page—it's not limited to one particular element or type of element. For example, I could apply my `cool` class to paragraphs or headings or even to table cells, wherever I wanted the text to look "cool."

Classes give you a very handy way to define different styles and apply them only when you want them. Incidentally, you can call your classes whatever you want, although it's best to use a name that you can remember easily and that will be clear to anyone else who may need to maintain the pages later. Keep the names short and don't use more than one word (you can't have spaces in the name).

An id functions in much the same way that a class does, but it gives a selector a unique name, which is useful in applications such as CSS positioning. Unlike a class, **you can use an id only once on a page** because it gives an element a unique name. If you're likely to use a style more than once on a page, use a class rather than an id.

Like a class, you define an id in a CSS style rule. Again, the only difference in the format comes with the selector: the id name must begin with a pound sign. The format for an id defnition is this:

```
#idname{property:value; property:value;}
```

A sample id rule might look like this:

```
#intro{font-size:large;}
```

Like classes, you can use ids with any HTML tag. To use an id with an HTML tag, you insert the id attribute in the tag, like this: `<p id="intro">`.

Here's a simple HTML page with an internal style sheet:

```
<!DOCTYPE HTML PUBLIC "-//W3C//DTD HTML 4.01
    Transitional//EN" "http://www.w3.org/TR/1999/
    REC-html401-19991224/loose.dtd">
<html>
<head>
    <title>CSS Basics</title>
    <style type="text/css">
body{margin-left: 50px;}
p{color: blue; font-family: verdana, helvetica,
    sans-serif; font-size: medium;}
h1{color: red; font-family: verdana, helvetica,
    sans-serif; font-size: large;}
</style>
</head>
<body>
<h1>A Basic Page</h1>
<p>
Cascading Style Sheets were developed by Hakon Lie
and Bert Bos, working for the World Wide Web
Consortium. The idea was to separate presentation—the
```

```
way the page looks—from structure—the way the page is
organized.
</p>
<p>
Using Cascading Style Sheets, you separate the style
elements of the page (the font, size, color, spacing,
etc.) and put them in a style section. Then if you
decide later that you want to make a change, you have
to deal with only one place, rather than hunting down
all the places where you used the old FONT and
BASEFONT tags.
</p>
<p>
Later on in this book, we'll cover the various prop-
erties available with Cascading Style Sheets, but
this will do for an introduction.
</p>
</body>
</html>
```

In a browser it will look like Figure 3–1.

Other Selectors

The easiest way to create a selector is to use the name of the element (e.g., p or h1), as we just did. But this isn't the only kind of selector you can use. For example, if you want to apply the same style to several different elements, you can use a list of selectors, separated by commas, like this:

h1,h2,h3,h4{color:red;}

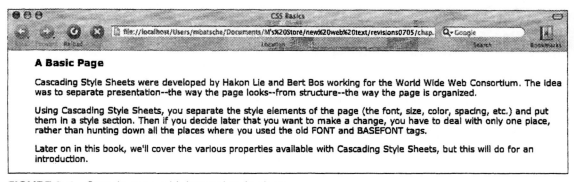

FIGURE 3–1 Sample page with internal style sheet

You can also identify selectors according to their parents or even their grandparents. For example, if you wanted to color all the paragraphs red within a division that has the id intro, you could use this selector:

```
#intro p{color:red}
```

The style rule here says that all the paragraphs that have a parent with the id intro (i.e., all paragraphs included within an element with the id intro) will be red. The general rule for context selectors is that elements are selected depending on their ancestors; an ancestor is any element that contains the element you're using as a selector.

To use an ancestor in a selector, type the name of the element that is the ancestor—that is, the element that contains the elements to which you want to apply the styles—followed by a space. Then type the name of the element to which you want to apply the styles, and complete the style rule.

You can also use part of an element—that is, the first line or the first letter—as a selector. To use the first line as a selector, type the name of the element whose first line you want to format, followed by a colon. Then type first-line to select the first line of the element, and type the rest of the style rule, that is, the styles you want applied to the first line. The result looks like this:

```
p:first-line{color:red;}
```

To use the first letter as a selector, type the name of the element whose first letter you want to format, followed by a colon. Then type first-letter to select the first letter of the element, and type the rest of the style rule, that is, the styles you want applied to the first letter. The result looks like this:

```
p:first-letter{color:red;font-size:x-large;}
```

CREATING A STYLE SHEET

Now let's return to the HTML page that we created in the previous chapter; if you'll look back at Figure 2–1, you'll see that it's not very appealing visually. However, there are several style properties that can make the page more effective. Here's the current page code:

```
<!DOCTYPE html PUBLIC "-//W3C//DTD XHTML 1.0
    Transitional//EN" "http://www.w3.org/TR/xhtml1/
    DTD/xhtml1-transitional.dtd">
<html xmlns="http://www.w3.org/1999/xhtml">
<head>
    <title>HTML Basics</title>
</head>
<body>
<h1>HTML Basics</h1>
```

```
<p>
All HTML tags are written between less-than and
greater-than signs (angle brackets). Many HTML tags
are container tags, that is, they have both an open-
ing and a closing tag. All the content that appears
between these tags will be affected by them. Think of
the opening tag as a switch that turns on the effect
and the closing tag as a switch that turns off the
effect.
</p>
<img src="compwriter.gif" />
<p>
The opening and closing tags use exactly the same
word or symbol (e.g., h1), but the closing tag has a
forwardslash before the word or symbol (e.g., /h1) in
order to differentiate it from the opening tag. If
you're using one of these container tags, you must
include the closing tag as well as the opening tag.
If you forget and leave off the close tag, your text
may not show up in the browser, or it may not have
the style you expect it to have because you didn't
turn the effect off after you turned it on. In HTML
some of these close tags were optional; in XHTML none
of them are. Since all Web pages will eventually be
written in XHTML, it's best to get into the habit of
including close tags for all container tags.
</p>
<a href="http://www.nypl.org/styleguide/">The New
York Public Library's XHTML Style Guide</a>
</body>
</html>
```

We're going to use an internal style sheet to add the styles, so we'll need to add the `<style></style>` tags to the head section, like this:

```
<head>
    <title>HTML Basics</title>
    <style type="text/css">
    </style>
</head>
```

EXERCISE: ADD AN INTERNAL STYLE SHEET

First, open the HTML page that you created in the previous chapter so that you can add some styles. After `</title>` but before `</head>`, type `<style type ="text/css"></style>`. You'll be adding style rules between these two tags.

Now that we have an internal style sheet set up, we can add some style rules to fix the page.

Margins

At the moment the lines of text on the page run almost from one side of the browser window to the other. As you'll discover in chapter 4, long text lines are harder to read. So let's shorten them by adding a margin of 50 pixels all around the page. We'll use body as the selector so that the margin will be applied to everything on the page. The style rule looks like this:

```
body{margin:50px;}
```

The internal style sheet now looks like this:

```
<head>
    <title>HTML Basics</title>
    <style type="text/css">
        body{margin:50px;}
    </style>
</head>
```

EXERCISE: ADD A MARGIN

Add a margin to the internal style sheet on your page. After <style type="text/css"> but before </style>, type body{margin:50px;}. You can use a different number of pixels for your margin if you'd like to experiment.

Fonts and Text

Next we'll use some of the CSS font properties to make changes in the appearance of the text. First, we'll change the heading font from serif to sans serif, and then we'll make it a little smaller. With h1 as the selector, the style rule looks like this:

```
h1{font-family: verdana, helvetica, sans-serif; font-size: large;}
```

In this rule I'm giving the browser a list of fonts for the main heading: Verdana is my first choice, but if the computer doesn't have Verdana loaded in the font selections, then the browser will look for Helvetica. If neither Verdana nor Helvetica is present, the browser will take the first sans serif font it comes to. We'll talk more about choosing fonts in chapter 4. The rest of the style rule sets the size of the font.

Now I'll fix the paragraphs. I'm not going to change the font (the combination of sans serif headings and serif text is a very familiar one), but I want to put some space between the text lines. For that I'll use another property: line-height. The style rule looks like this:

```
p{line-height: 1.5;}
```

In this rule I've told the browser to put a space and a half between the text lines, rather than the default single spacing. The style sheet now looks like this:

```
<head>
    <title>HTML Basics</title>
    <style type="text/css">
        body{margin:50px;}
        h1{font-family: verdana, helvetica, sans-
serif;font-size: large;}
        p{line-height: 1.5;}
    </style>
</head>
```

I'm going to add one more property to the rule for the heading. The default alignment for headings is on the left margin, but I'd like the heading to be centered. That property is text-align. The style rule now looks like this:

```
h1{font-family: verdana, helvetica, sans-serif; font-
size: large; text-align: center;}
```

And the style sheet now looks like this:

```
<head>
    <title>HTML Basics</title>
    <style type="text/css">
        body{margin:50px;}
        h1{font-family: verdana, helvetica, sans-
serif; font-size: large; text-align: center;}
        p{line-height: 1.5;}
    </style>
</head>
```

Let's see what the page looks like with the styles added (See Figure 3–2).

There are still things that need to be done to this page: the spacing between the graphic and the text leaves an unattractive gap, and the link could use some

---

EXERCISE: ADD STYLES TO YOUR TEXT

In the style section on your Web page (after body{margin:50px;}), type h1{font-family: and list the fonts you'd like to use for your headings, starting with your first choice and ending with a generic (use either serif or sans-serif for now). Remember that these fonts must be loaded on the computer for the browser to read them. Separate the fonts on your list with commas, and end with a semicolon. Then type font-size: large; text-align:center;}. Finally, add another style rule: p{line-height:1.5;}.

HTML Basics

All HTML tags are written between less-than and greater-than signs (angle brackets). Many HTML tags are container tags, that is, they have both an opening and a closing tag. All the content that appears between these tags will be affected by them. Think of the opening tag as a switch that turns on the effect and the closing tag as a switch that turns off the effect.

The opening and closing tags use exactly the same word or symbol (e.g., h1), but the closing tag has a forward slash before the word or symbol (e.g., /h1) in order to differentiate it from the opening tag. If you're using one of these container tags, you must include the closing tag as well as the opening tag. If you forget and leave off the close tag, your text may not show up in the browser, or it may not have the style you expect it to have because you didn't turn the effect off after you turned it on. In HTML some of these close tags were optional; in XHTML none of them are. Since all Web pages will eventually be written in XHTML, it's best to get into the habit of including close tags for all container tags.

The New York Public Library's XHTML Style Guide

FIGURE 3–2   Sample page with styles

styles as well, but this is a good start. We'll get more deeply into HTML and CSS applied to text in the next chapter.

CREATING BASIC PAGES

You now have enough HTML and CSS information to create basic page. As you can see, Web code isn't particularly difficult, once you have the basic formats down. Pages like the example in this chapter don't have much design, but they're adequate for simple content.

PROCESS FOR CREATING A BASIC INTERNAL STYLE SHEET

Use the following steps to add an internal style sheet to a basic XHTML page:

1. After `</title>` but before `</head>`, type `<style type="text/css"></style>`.

2. After `<style>` type the following rule to add a margin on all sides of your page:

```
body{margin:50px;}
```

3. After `body{margin:50px;}` type the following rule to add styles to your heading:

`h1{font-family: verdana, helvetica, sans-serif; font-size: large; text-align: center;}`

4. After `h1{font-family: verdana, helvetica, sans-serif; font-size: large; text-align: center;}`, type the following rule to add space between the text lines of your paragraphs:

`p{line-height:1.5;}`

5. Save your file with the `.html` extension.

ASSIGNMENT: WEB SITE

Continue working on the pages you created using the basic HTML tags, and add styles to the pages. Use either an internal style sheet as we did in the example in this chapter or an external style sheet that you'll apply to the entire site.

CHAPTER SUMMARY

Web Standards

❑ Adherence to Web standards means using only the tags included in HTML 4.01 or XHTML, along with CSS.

❑ HTML 4.01 and XHTML 1.0 deprecate several tags used in earlier versions of HTML. Although these tags may still be read by current browsers, newer browser versions may no longer recognize them.

❑ Web designers should avoid both deprecated tags and attributes and browser-specific tags and attributes.

CSS

❑ CSS allows Web designers to separate the structure of their pages (done through HTML) from the presentation or appearance of the pages.

❑ CSS can be applied through an external style sheet, which applies to all the pages that are linked to the style sheet file; through an internal style sheet, which is placed in the head section of a page; or through a `style` attribute placed within a particular HTML tag in the body section.

CSS Basics

❑ CSS rules consist of a selector followed by a property (or properties) and value(s)

❑ CSS classes are special combinations of styles that are defined by the designer and applied to several selectors on a page. CSS ids are special combinations of styles that are defined by the designer and applied to only one selector per page (used mainly in CSS positioning).

❑ Selectors can also be lists of elements or can be based on context.

SOURCES

Berners-Lee, T. 2000. *Weaving the Web*. New York: HarperBusiness.

Zeldman, J. 2003. *Designing with Web standards*. Indianapolis, IN: New Riders.

Working with Text: HTML Tags

Chapter
4

TEXT AND HTML

Because HTML began as a way for scientists to exchange textual information, such as reports and minutes of meetings, it's not surprising that HTML contains a full selection of text tags. These are structural tags, indicating the structural role each text segment plays on your Web page. CSS includes an equally broad spectrum of properties for adding presentation styles to your text passages. In this chapter we'll discuss some of the basic HTML text tags that you can use, along with the tags for paragraphs and headings that you learned in chapter 2. First, however, we need to discuss some principles of writing text for the Web.

WRITING TEXT FOR THE WEB

If there's one characteristic that distinguishes effective Web text, it's brevity. Steve Krug (2000) is exaggerating only slightly when he argues, "Get rid of half the words on each page, then get rid of half of what's left" (45). Research by Jakob Nielsen (2000) indicates that we read up to 25 percent more slowly from a computer screen than from print (102). Thus, it makes sense to try to cut back on the amount of text that users have to get through and to write in a style that makes it possible to read that text easily.

Jonathan and Lisa Price (2002) argue that, above all, a Web writer should "write like a human being" (84). This means, once again, reducing the amount of text. The Prices advise reducing by 50 percent any text that was originally written for paper when adapting it to the Web. They further recommend eliminating "words that don't add anything to what you have already said, unnecessary details . . . and phrases that tell the readers something they already learned earlier" (88). In addition, the Prices suggest that Web paragraphs should be shortened to allow readers to skim; their ideal is two to three lines per paragraph. And they argue against hyperbole and promotional material, which most readers will skip. They suggest creating separate pages for supplemental material that only some readers will need—pages that can then be linked to the main content.

Text Guidelines

Organization. Since most Web users scan text, It's a good idea to arrange your text to make scanning easier. Two to three levels of heads and subheads that clearly identify the content will help. Nielsen advises using inverted-pyramid-style organization for text blocks; this is the organization used in print news stories, beginning with the central point of the text block and following with major facts, then filling in with less vital information, which can be cut if necessary. To use inverted-pyramid organization in your text blocks, put a statement of the block's subject (the equivalent of the topic sentence in a paragraph) at the beginning of the block, and follow it with important supporting information. Any further information can be placed later in the text block, allowing skimming readers to pick up the most important points without losing vital information as they skip to the next block.

Tone. The tone of the "voice" in your writing should match the tone of your site as a whole. If you're dealing with a serious topic (e.g., a nonprofit charity or a social/political appeal), your tone should match the level of seriousness. On the other hand, because the Web is frequently more personal than print, many sites use an informal style. If you want to address your reader directly, your tone should be more light.

First and Second Person. Many Web sites are written in first person, even using first person plural to represent an organization. In this case the user is addressed directly in second person. This use of first and second person is typical of informal business and technical writing and may help to keep the reader involved. On the other hand, unnecessary passive voice and convoluted sentence structure can be difficult to follow, especially on the computer screen. Keep your prose active and direct.

Humor. Catching a viewer's attention on the Web can be a struggle, and sometimes humor can be an effective technique for doing so. Humor can help to explain complex topics and may make your site more memorable. Unfortunately, it can also mark your site as stupid or even offensive. Use a light touch, and be sure to check the reactions of other readers by having them read your pages before you mount them.

Language Technicalities. You should always use a reference resource for questions of grammar and mechanics. *The Chicago Manual of Style* and *The Associated Press Stylebook and Libel Manual* can both provide rules for questions of punctuation, capitalization, grammar, and usage. A good dictionary can help with spelling. In addition, you should always run your text through a spell check in a Web design program or HTML editor.

You should read your text carefully for errors in punctuation, subject-verb agreement, pronoun reference, and other mechanical matters. If possible, have someone else read over your text strictly for mechanical errors. Frequently, it's difficult to catch textual errors when you're creating content and trying to be aware of coding and design. Another pair of eyes can make your job much easier.

USING HTML/XHTML TEXT TAGS

Now that you have some idea of the requirements for writing online text, you can use HTML/XHTML to structure it. In chapter 5 we'll discuss using CSS to design the text presentation.

Paragraphs and Line Breaks

We've already covered the paragraph tag (`<p></p>`) in chapter 2. Remember that paragraphs in basic HTML are not indented; they're separated by a line of vertical space. (In the next chapter we'll discuss a way to indent them using CSS.)

Occasionally, you may want to create a new line of text without having the extra vertical space you get when you create a new paragraph or use another block-level tag like `h1`. For example, if you were writing a poem in HTML, you might not want extra space between the lines. In that case you could use the line break tag `<br>` (HTML) or `<br />` (XHTML) to start a new line without adding any extra space between the lines. If you insert the line break tag at the end of the line, the next text line will begin on a new line without extra space between the lines. Since `<br>` is an empty tag—that is, not a container tag—you don't need to use `</br>`.

Sometimes novice Web writers try to create extra space on a page by using multiple line breaks, like this: `<br><br><br><br><br><br>`. Unfortunately, this doesn't work; some browsers will simply collapse all those line breaks into a single line break. We'll discuss more effective ways to create white space with CSS later on. Here's a simple XHTML page that includes some line breaks:

```
<!DOCTYPE html PUBLIC "-//W3C//DTD XHTML 1.0
    Transitional//EN" "http://www.w3.org/
    TR/xhtml1/DTD/xhtml1-transtional.dtd">
<html xmlns="http://www.w3.org/1999/xhtml">
<head>
    <title>HTML Text Tags</title>
    <style type="text/css">
    body{margin:75px;}
h1{font-family: verdana, helvetica, sans-serif;
    font-size: large; text-align: center;}
    p{line-height: 1.5;}
</style>
</head>
<body>
<h1>HTML Text Tags</h1>
<p>
```

```
HTML text tags indicate the structure of HTML pages;
that is, they indicate what role a text segment is
playing in the page organization. Using these text
tags, you can indicate which text segments are
headings, subheadings, lists, quotes, and paragraphs.
</p>
<p>
HTML line-break tags can be used for poetry or other
situations in which you want text to appear on a
separate line without any extra vertical space. Using
a line break looks like this:<br />
Mary had a little lamb<br />
Its fleece was white as snow.
</p>
</body>
</html>
```

In a browser this page would look like Figure 4–1.

Headings

In HTML there are six levels of headings. You've already seen the highest level, `<h1></h1>`. If you want to create a smaller-sized heading, type `<h2>` or `<h3>` (and so on) instead of `<h1>` and `</h2>` or `</h3>` instead of `</h1>`. Because headings are preformatted by the browser, they'll be bolded and sized according to the type of computer you're using to view the page (type looks smaller on a high-resolution monitor) and the type of browser you're using to look at them, unless you use CSS to change them. They will also be placed against the left margin unless you use CSS to change the alignment. Remember, headings (like paragraphs) are block-level elements (i.e., they have an automatic line break afterward), so you don't need to follow the `</h1>` with `<br>` or `<br />`.

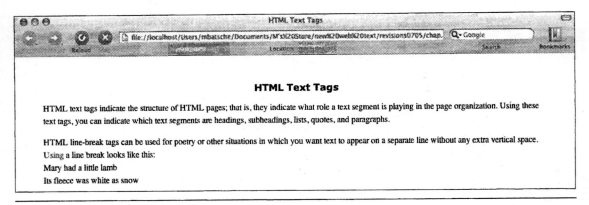

FIGURE 4–1 A page with line breaks

EXERCISE: LINE BREAKS

Check the text in your content inventory. Do any of the pages have text segments that need line breaks? If so, mark where the line break tags should be placed. If you're ready to write the pages, insert the line-break tags in your code.

Block Quotes

You can use the HTML `<blockquote></blockquote>` tag to indent a section of text, setting it off as you would indent a long quotation in a paper. Don't rely on block quotes to create page layout: they're not presentation elements. They're strictly for indenting sections of text and are most useful for their original purpose, creating block quotes, particularly if you use the `<cite></cite>` tags, which we'll discuss later in this chapter. Here's the earlier page with an added block quote:

```
<!DOCTYPE html PUBLIC "-//W3C//DTD XHTML 1.0
     Transitional//EN" "http://www.w3.org/TR/
     xhtml1/DTD/xhtml1-transitional.dtd">
<html xmlns="http://www.w3.org/1999/xhtml">
<head>
    <title>HTML Text Tags</title>
    <style type="text/css">
    body{margin:50px;}
h1{font-family: verdana, helvetica, sans-serif;
    font-size: large; text-align: center;}
    p{line-height: 1.5;}
</style>
</head>
<body>
<h1>HTML Text Tags</h1>
<p>
HTML text tags indicate the structure of HTML pages;
that is, they indicate what role a text segment is
playing in the page organization. Using these text
tags, you can indicate which text segments are
headings, subheadings, lists, quotes, and paragraphs.
</p>
<p>David Siegel, a Web designer in the 1990s, once
said this of the basic Web style in early HTML: </p>
<blockquote>
"First generation sites were designed by technical
people. Some sites had headline banners and were well
organized; most had edge-to-edge text that ran on for
pages, separated by meaningless blank lines. At best,
```

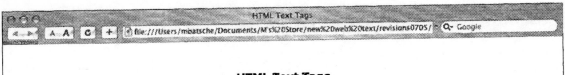

FIGURE 4–2 A page with a block quote

```
they looked like slide presentations shown on a
concrete wall."
</blockquote>
</body>
</html>
```

In a browser it would look like Figure 4–2 (with default paragraph spacing).

EXERCISE: BLOCK QUOTES

Check the text in your content inventory. Do any of the pages have long quotes? If so, mark them to be done as block quotes. If you're already writing your HTML pages, place the quotes within the `<blockquote></blockquote>` tags.

Lists

Lists are helpful for presenting information quickly and making it easy to skim, two major requirements for Web writing. Another standard use for lists is in the presentation of links. HTML has three types of lists: ordered, unordered, and definition.

Ordered Lists. Ordered lists use numbers. You should use an ordered list when the items you're listing have a particular order you need to maintain, like a series of steps in a procedure. Ordered lists begin and end with container tags: `<ol></ol>`. Then within these outer containers, you use a series of list item tags `<li></li>` to mark each item on the list. When you look at the list in the browser, the items will be numbered in order. Here's the XHTML page again, with an ordered list:

```
<!DOCTYPE html PUBLIC "-//W3C//DTD XHTML 1.0
       Transitional//EN" "http://www.w3.org/TR/
       xhtml1/DTD/xhtml1-transitional.dtd">
```

```
<html xmlns="http://www.w3.org/1999/xhtml">
<head>
    <title>HTML Text Tags</title>
    <style type="text/css">
    body{margin:50px;}
h1{font-family: verdana, helvetica, sans-serif;
    font-size: large; text-align: center;}
    p{line-height: 1.5;}
</style>
</head>
<body>
<h1>HTML Text Tags</h1>
<p>
HTML text tags indicate the structure of HTML
pages; that is, they indicate what role a text
segment is playing in the page organization.
Using these text tags, you can indicate which
text segments are headings, subheadings, lists,
quotes, and paragraphs.
</p>
<p>
The head section of this XHTML page contains the
following items:
</p>
<ol>
    <li>DOCTYPE</li>
    <li>HTML namespace</li>
    <li>Head tag</li>
    <li>Title tags and page title</li>
    <li>Internal style sheet</li>
</ol>
</body>
</html>
```

In a browser it would look like Figure 4–3.

Unordered Lists. If the order of the items in your list doesn't matter (e.g., a list of tools needed for a procedure), use an unordered list. The HTML for an unordered list is very similar to that used for ordered lists. The container tags are <ul></ul>, and within the container you use the same list item tags, <li></li>. Here's a version of the XHTML page with an unordered list:

```
<!DOCTYPE html PUBLIC "-//W3C//DTD XHTML 1.0
    Transitional//EN" "http://www.w3.org/TR/
    xhtml1/DTD/xhtml1-transitional.dtd">
```

FIGURE 4–3 A page with an ordered list

```
<html xmlns="http://www.w3.org/1999/xhtml">
<head>
    <title>HTML Text Tags</title>
    <style type="text/css">
    body{margin:50px;}
h1{font-family: verdana, helvetica, sans-serif; font-
    size: large; text-align: center;}
    p{line-height: 1.5;}
</style>
</head>
<body>
<h1>HTML Text Tags</h1>
<p>
HTML text tags indicate the structure of HTML
pages; that is, they indicate what role a text segment
is playing in the page organization. Using these text
tags, you can indicate which text segments are head-
ings, subheadings, lists, quotes, and paragraphs.
</p>
<p>
Unordered lists can be used for many things:
</p>
<ul>
    <li>Link lists</li>
    <li>Summaries</li>
    <li>Materials lists</li>
<li>Anything for which the order of the list items
    doesn't matter</li>
</ul>
```

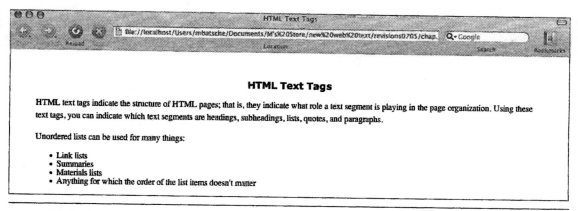

FIGURE 4–4   A page with an unordered list

```
</body>
</html>
```

In a browser it would look like Figure 4–4.

Definition Lists. Definition lists use a term-and-definition order—a term, placed at the left margin, and an indented definition. The list container is `<dl></dl>`, the tags for the definition term are `<dt></dt>`, and the definition tags are `<dd></dd>`. The XHTML version looks like this:

```
<!DOCTYPE html PUBLIC "-//W3C//DTD XHTML 1.0
     Transitional//EN" "http://www.w3.org/TR/
     xhtml1/DTD/xhtml1-transitional.dtd">
<html xmlns="http://www.w3.org/1999/xhtml">
<head>
    <title>HTML Text Tags</title>
    <style type="text/css">
    body{margin:50px;}
h1{font-family: verdana, helvetica, sans-serif;
    font-size: large; text-align: center;}
    p{line-height: 1.5;}
</style>
</head>
<body>
<h1>HTML Text Tags</h1>
<p>
HTML text tags indicate the structure of HTML pages;
that is, they indicate what role a text segment is
playing in the page organization. Using these text
tags, you can indicate which text segments are
headings, subheadings, lists, quotes, and paragraphs.
</p>
```

```
<p>
Here are some definitions of HTML terms:
</p>
<dl>
    <dt>Tag</dt>
        <dd>An instruction to browsers that specifies
the structure of Web content. </dd>
    <dt>Attribute</dt>
        <dd>A word or symbol placed within a tag that
affects the appearance or behavior of an
            element on a page.</dd>
    <dt>Block-level tag</dt>
        <dd>An HTML element that begins on a separate
text line and frequently has extra space
            inserted before it.</dd>
    <dt>In-line tag</dt>
        <dd>An HTML element that neither begins on
a separate text line nor inserts extra space.
It affects the appearance and behavior of an
element within a line of text.</dd>
</dl>
</body>
</html>
```

In a browser it would look like Figure 4–5.

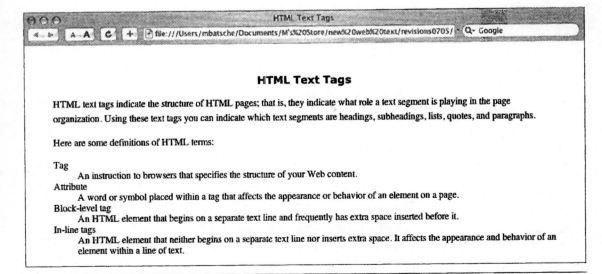

FIGURE 4–5   A page with a definition list

SIDEBAR: LIST TIPS

❑ Lists are block-level elements, so you don't need to insert a line break (`<br>` or `<br/>`) at the end of a list item (`</li>`).

❑ To indent another set of items below a list item (e.g., to use outline style), you can use nested lists, placing one list inside another. Just be sure to close all the tags in your nested list before you close the tags in the container list.

❑ Keep the text in your list items short.

EXERCISE: LISTS

You'll undoubtedly have some sections of your content that would work best as ordered or unordered lists, and you may have terms that need to be defined. Look through your content inventory for possible lists, and mark those that you find as either ordered or unordered. If you've begun your HTML pages, try formatting the lists as either ordered lists (if the order matters) or unordered lists (if the order doesn't matter). If you have several terms to define, try creating a glossary, using a definition list format.

PHRASE ELEMENTS

The W3C has another class of HTML text tags that are referred to as phrase elements. According to the W3C specifications, phrase elements "add structural information to text fragments." Some of the more familiar phrase elements include `<em></em>`, `<strong></strong>`, `<cite></cite>`, `<abbr></abbr>`, and `<acronym></acronym>`.

`<em>` and `<strong>`. Perhaps the most familiar phrase element tags are `<em>`, which stands for *emphasis,* and `<strong>`, which stands for *strong emphasis.* In most modern browsers, text between `<em>` and `</em>` will be rendered in italics, whereas text between `<strong>` and `</strong>` will be rendered in boldface. Older versions of HTML also offered an italic tag, `<i></i>`, and a boldface tag, `<b></b>`. However, like most presentational tags, `<i>` and `<b>` have been dropped in favor of the CSS properties `font-style` and `font-weight`. However, `<em>` and `<strong>` remain in use because they indicate text to be emphasized rather than text that has a particular visual style.

For example, say you wanted to include this text on your Web page:

> Use HTML tags to indicate the **structure** of your page and CSS properties to indicate the **presentation**.

The words *structure* and *presentation* are being boldfaced to emphasize them rather than to add a particular visual style. If a browser chose to render *structure* and *presentation* in italics, the essential meaning of the passage would not be changed—the two words would still be emphasized. On the other hand, if you chose to style all of your links in boldface, that would be a presentation style—you're not trying to emphasize the text, just giving it a particular visual effect. In that case you'd use the CSS property `font-weight` to render the boldface for your links.

Another way to think about this distinction is to remember that screen readers that read Web pages for visually disabled users will emphasize words that are tagged with `<em>` and will emphasize words more strongly that are tagged with `<strong>`. On the other hand, text that has been styled with `font-weight` and `font-style` will receive no special emphasis from a screen reader. When you're deciding whether to use `<em>` and `<strong>` or `font-style` and `font-weight`, consider whether the words would be particularly emphasized if you read them out loud or whether their styling is strictly visual.

`<cite></cite>`. If you need to make reference to a source citation—that is, an author or a publication title—you can use the `<cite></cite>` tag. Most browsers will render text included within these tags in italics, but you can create special styles for `<cite>` on your style sheet. Using the `<cite>` tag for references will make your page easier to search and easier to sort into databases because it will separate citations from other text that you want to emphasize or style with italics.

`<abbr></abbr>` *and* ***`<acronym></acronym>`***. Technically, `<abbr>` and `<acronym>` refer to different things: `<abbr>` stands for *abbreviation* (e.g., HTTP, HTML, UTSA) whereas `<acronym>` refers to acronyms (e.g., NASA, NATO, laser). One way of remembering the distinction is that screen readers will spell out abbreviations, while pronouncing acronyms as if they were words. Note that `abbr` is also used as an attribute with tables (see appendix 1). Both `<abbr>` and `<acronym>` usually include a `title` attribute in order to indicate the full title that's being abbreviated, like this:

```
<acronym title="National Aeronautics and Space
Administration"> NASA</acronym>
<abbr title= "University of Texas at San Antonio">
UTSA </abbr>
```

Some browsers style abbreviations and acronyms with a dotted bottom border to alert users that they can move their cursors across the text to see the definition in a tool-tip box. Unfortunately, according to Dan Cederholm (2004), Internet Explorer for Windows does not support this styling for `<abbr>`, only for `<acronym>`.

Here's a page using phrase elements:

```
<!DOCTYPE html PUBLIC "-//W3C//DTD XHTML 1.0
    Transitional//EN" "http://www.w3.org/TR/
    xhtml1/DTD/xhtml1-transitional.dtd">
<html xmlns="http://www.w3.org/1999/xhtml">
<head>
    <title>HTML Text Tags</title>
    <style type="text/css">
```

```
    body{margin:50px;}
    h1{font-family: verdana, helvetica, sans-serif;
    font-size: large; text-align: center;}
    p{line-height: 1.5;}
</style>
</head>
<body>
<h1>HTML Text Tags</h1>
<p>
HTML text tags indicate the structure of HTML pages;
that is, they indicate what role a text segment is
playing in the page organization. Using these text
tags, you can indicate which text segments are
headings, subheadings, lists, quotes, and paragraphs.
</p>
<p>
Phrase elements "add structural information to text
fragments," according to the <abbr title="World Wide
Web Consortium">W3C</abbr>. These elements may seem
to be adding visual style to your pages, but in fact
they're adding <em>structural</em> information. The
text that's placed within phrase element tags will be
treated differently by screen readers from text that
has purely visual styles applied. The tags indicate
something about the <strong>nature of the text
itself</strong> beyond its appearance.
</p>
<p>
Dan Cederholm's excellent book, <cite>Web Standards
Solutions: The Markup and Style Handbook</cite>, has
an entire chapter on phrase elements, with styling
suggestions that can be added to your style sheets.
</p>
</body>
</html>
```

In a browser it would look like Figure 4–6.

EXERCISE: PHRASE ELEMENTS

Since phrase elements offer code for specialized situations, you'll have few items in your content inventory that seem to require them. However, it may help to go through your content now and mark information that will use phrase elements when your pages are set up. Mark source citations, abbreviations, acronyms, and any words that may need special emphasis (i.e., words that may already be boldfaced or in italics).

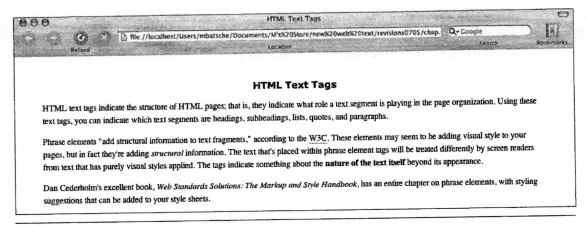

FIGURE 4–6   A page with phrase elements

USING HTML TEXT TAGS

You now know enough HTML to create a Web page that can communicate textual information. Some Web page authors don't go much further than this; it suits their purposes.

If you haven't looked at your HTML page in a browser, do it now. At this point you may not be too impressed by what you see when you check your page because few effective pages on the Web are text only. To make an interesting Web page, you need to use more than just the basics. That's where the text properties in CSS come in, which we'll cover in the next chapter.

PROCESS FOR CREATING HTML TEXT TAGS

Use the following steps to code your text pages:

1. Make sure that the text for your Web pages follows the guidelines for Web writing: short, direct, and organized for easy scanning.
2. Create paragraphs by using the `<p></p>` tags. To start a new line without any additional space between the lines, use the line break tag `<br>` (HTML) or `<br/>` (XHTML) at the end of the line where you want the break to occur.
3. Create the highest level of heading by using `<h1></h1>` tags. For lower-level headings, use `<h2></h2>`, `<h3></h3>`, `<h4></h4>`, `<h5></h5>`, and `<h6></h6>`.
4. For quotes longer than a single line, use `<blockquote></blockquote>`.
5. If list items must be presented in a particular order, use `<ol><li></li></ol>`.
6. If the order of items in a list is unimportant, use `<ul><li></li></ul>`.
7. For a glossary of terms, use `<dl><dt></dt><dd></dd></dl>`.

8. If the text includes words and phrases that should be emphasized, use `<em></em>` or `<strong></strong>`.
9. If the text includes source citations, indicate them with `<cite></cite>`.
10. If the text includes abbreviations, indicate them with `<abbr></abbr>`. If it uses acronyms, indicate them with `<acronym></acronym>`.

CHAPTER SUMMARY

Writing Text

❑ Text written for the Web should be brief, subdivided with headings and sub-heads for easy scanning, and organized in an inverted-pyramid style.

❑ Text should be written in a tone matching the seriousness of the site; first and second person can be used; and humor can be used judiciously.

❑ All text posted on the Web should be double-checked for mechanical and grammatical accuracy.

HTML Text Tags

❑ HTML includes a full range of text tags to indicate the structure of a document, including paragraphs and line breaks, headings, block quotes, lists, and phrase elements.

❑ Web designers should always check their pages in a browser, preferably in more than one browser.

HTML/XHTML TAGS IN THIS CHAPTER

TAG	EFFECT
`<!DOCTYPE HTML PUBLIC "-//W3C//DTD HTML 4.01 Transitional//EN" "http://www.w3.org/TR/html4/loose.dtd">` `<html></html>` (HTML)	Indicates variety of HTML or XHTML being used: Transitional, Strict, or Frames
`<html xmlns="http://www.w3.org/1999/xhtml"></html>` (XHTML)	Indicates that page code is HTML
`<head></head>`	Establishes head section of a page
`<title></title>`	Indicates page title
`<link rel="stylesheet" type="text/css" href="mystyles.css">` (HTML)	Links page to external style sheet
`<link rel="stylesheet" type="text/css" href="mystyles.css" />` (XHTML) `<style type="text/css"> </style>`	Inserts an internal style sheet into the head section
`@import url(styles.css)`	Links page to external style sheet (placed between `<style>` and `</style>`)

Continued

Continued

TAG	EFFECT
`<body></body>`	Establishes body section of a page
`<p></p>`	Creates a paragraph
` ` (HTML)	Creates a line break
` ` (XHTML)	
`<h1></h1>  <h3></h3>  <h5></h5>` `<h2></h2>  <h4></h4>  <h6></h6>`	Creates a heading
`<blockquote></blockquote>`	Creates an indented block quote
`<ol></ol>`	Creates an ordered (numbered) list
`<ul></ul>`	Creates an unordered (bulleted) list
`<li></li>`	Creates a list item in either an ordered or unordered list
`<dl></dl>`	Creates a definition list
`<dt></dt>`	Creates a term to be defined in a definition list
`<dd></dd>`	Creates a definition in a definition list
`<em></em>`	Places emphasis on a word or phrase
`<strong></strong>`	Places strong emphasis on a word or phrase
`<cite></cite>`	Indicates a reference citation
`<abbr></abbr>`	Indicates an abbreviation
`<acronym></acronym>`	Indicates an acronym

HTML/XHTML ATTRIBUTES IN THIS CHAPTER

ATTRIBUTE	VALUE	EFFECT
title	Complete version of abbreviation or acronym	Provides full version of abbreviation or acronym when cursor is passed across it

SOURCES

Cederholm, D. 2004. *Web standards solutions: The markup and style handbook.* Berkeley, CA: Friends of Ed/Apress.

Krug, S. 2000. *Don't make me think: A commonsense approach to Web usability.* Indianapolis, IN: Que.

Nielsen, J. 2000. *Designing Web usability.* Indianapolis, IN: New Riders.

Price, J., and L. Price. 2002. *Hot text: Web writing that works.* Indianapolis, IN: New Riders.

Working with Text: Cascading Style Sheets

Chapter

5

CSS AND TEXT

As I've said before, HTML should be used to structure Web pages, and CSS should be used to give them visual presentation. All Web sites have (or should have) a visual element; in fact, the visual presentation of your site may be what makes a user stick around—or leave hurriedly! In this chapter we'll go over the properties included in CSS that apply particularly to text, although some possible text properties, like color, will be covered in greater detail in chapter 7. Before we look at the properties, however, let's consider some basic design principles that apply particularly to text pages.

DESIGNING TEXT

Your primary concern in designing text sections is legibility. We'll discuss legible color combinations in chapter 7; but at this point it's enough to say that you need a high degree of contrast between text color and background color. Dark text on a light background is the most legible combination, but other combinations can be used for shorter text segments.

Justification

Left justification is the best choice for most text blocks because both centered and right-justified text are difficult to read in long text segments. Patrick Lynch and Sarah Horton (2001) argue that titles should also be left justified above left-justified text (122). Although centered titles are common with fully justified text, they can look unbalanced with a left-justified page and a ragged right margin.

Choosing Fonts

The choice of fonts for Web pages is more problematic. The fonts available for text are limited by the fonts loaded on a user's computer. Later in this chapter, we'll discuss which fonts are most likely to be found on various computer platforms; however, your choices will always be more limited than they would be for a printed document. Many designers recommend using typefaces designed for the screen, such as Microsoft's Verdana and Georgia, since they have been designed specifically for online use.

Whether you choose serif or sans serif type for your text font is largely a matter of taste. Some research suggests that sans serif type like Verdana is more readable at small sizes on lower-resolution screens. However, research has also found that most readers seem to prefer serif type for long text sections. It's best to judge whether a given font will be legible for long text segments by looking at them on the screen and having other readers give you their reactions.

At one time it was common for Web designers to create headings in a graphics program and then save them as GIF graphics in order to use special display fonts that would not be found on most users' computers. That particular technique has been discouraged in the Web Content Accessibility Guidelines adopted by the W3C. However, accessibility expert Joe Clark (2003) argues that these "pictures of text" can still be used with care. He advises, "If you use pictures of text, do so only for brief segments, as in navbars. Always use an alt text and, whenever possible, a title; both should reiterate the actual text" (227). Another alternative is to use a new technique called image replacement, which may be an effective way of creating accessible graphic headers. For more information on image replacement, see appendix 2.

Font Size

Like the choice of font, the choice of font size involves various considerations. Accessibility guidelines recommend using a relative font size so that readers can resize their type to accommodate vision problems. Pixels are the dots that make up the image displayed on a computer monitor. (*pixel* stands for "picture element"). Joe Clark (2003) suggests that "pixel (px) sizing is the most compatible with text-zoom features on today's browsers, and . . . any type less than nine pixels in height is unreadable" (223). However, Clark also points out that visually impaired users use screen magnification programs that will blow up the entire screen, and thus no particular type-size accommodation is necessary. Although using pixel sizes is generally reliable, Jeffrey Zeldman (2003) points out that Internet Explorer for Windows does not provide text zoom for type in pixel sizes, which will frustrate some users (245). In general, you can use a relative unit such as ems to size your type, or you can use the preset sizes that CSS provides (i.e., xx-small, x-small, small, medium, large, x-large, and xx-large), or you can use pixels. Just don't go below nine pixels for your smallest type.

Line Length

One problem that many Web sites share is lines of text that are too long for easy reading. Long text lines can cause eye strain and can also result in users' losing their place on the page. However, as Lynch and Horton (2001) point out, some users with larger monitors may not like having long pages with skinny text blocks, and users with vision disabilities may not appreciate a very short line length when they enlarge type (123–24).

One of the simplest ways to limit your line length is the method used in chapter 3—adding margins to the page. This method will be effective for simple page layouts like those we worked with in chapters 2 and 3, but more-complex page layouts will require more-complex solutions.

By using relative sizing for your text blocks, you can limit the length of your lines while allowing your users to set the page width to their own preferences. For example, you can set the width of your content to 80 percent of the browser window, which will have the effect of shortening the length of your text line. This relative width, however, will still accommodate different browser window sizes and monitor resolutions. If a fixed width for your text is necessary for some reason, Lynch and Horton (2001) recommend a maximum of 365 pixels, which will yield a line about fifty characters long, an average of nine to ten words (125).

CSS FONT PROPERTIES

The font properties in CSS were the first to be implemented by the major browsers, allowing you to set the font, the font size, the font weight, the font style, and the font color.

`font-family`

The `font-family` property allows you to select a particular font for your text. However, your page can be displayed only in a font that has been loaded on a reader's computer. So if you specify a font like Garamond and Garamond isn't loaded on your visitor's computer, your page will be displayed in a default font, like Times. CSS gets around this problem by allowing you to select more than one font and then list them in order of preference. So your CSS rule might look like this:

```
h2{font-family:garamond, palatino, georgia, times,
    serif;}
```

The browser will go down your font list in order until it finds a font loaded on the reader's computer that matches one of yours.

In this example, if the browser doesn't find Garamond on the reader's computer, it will look next for Palatino, then Georgia, then Times, and then a generic serif font. You should always end your list with a generic choice, as I did here. That allows the browser to try to find a font that looks roughly like the font you originally specified. The generic choices are serif, sans serif, cursive (i.e., script fonts like Zapf Chancery), fantasy (i.e., display fonts), or monospace (e.g., Courier). Cursive and fantasy may give you unexpected results because various cursive and display fonts look quite different and the browser will choose the first one it comes to on the user's computer. In fact, some browsers do not recognize cursive or fantasy at all. Another point to remember is that if you specify a font that has more than one word in its name, you must place it in quotation marks (e.g., "Century Gothic").

We used `font-family` to style the heading in chapter 3, but let's try it again with some different choices and an external style sheet. If we start with the styles we've already defined for the sample page, the external style sheet will look like this:

```
body{margin:50px;}
h1{font-family: verdana, helvetica, sans-serif; font-
    size: large;}
p{line-height: 1.5;}
```

SIDEBAR: COMMON FONTS

For the text part of your Web page, you'll want to use fonts that are available on most computers. The most widely supported fonts are Times, Arial, Helvetica, and Courier; these four will show up in some form on most computers. Beyond these four, however, things get a little more complicated. Windows and Macintosh have different default fonts, and different versions of Windows and the Mac OS also have different default fonts. Many Macintosh users have picked up additional Windows fonts through Microsoft programs like Internet Explorer; however, few Windows owners will have Macintosh fonts, like Monaco. Other software producers like Adobe and Macromedia include packages of fonts with some of their programs, which are available on many computers. However, since you can never be sure that a reader has any fonts beyond the standard ones, you should be careful to include one of those standards whenever you designate a font in CSS.

- ❑ **The following fonts are available on most Windows and Macintosh machines:** Arial, Times New Roman, Courier New, Verdana, Courier, Arial Black, Comic Sans MS, Trebuchet MS, Impact, Georgia, Tahoma, Arial Narrow, Helvetica, Times.
- ❑ **The following fonts are available on most Windows machines and some (although not all) Macs:** MS Sans Serif, Century Gothic, Garamond, Bookman Old Style, Book Antiqua, Haettenschweiler, and Monotype Corsiva.
- ❑ **The following fonts are available on most Macintosh machines and some (although not all) Windows machines:** Papyrus, Arial Rounded MT Bold, Brush Script MT, Andale Mono, Palatino, Gill Sans, Optima, Skia, Futura, Baskerville, Copperplate, Hoefler Text, Herculanum, Apple Chancery, Helvetica Neue, Techno, Zapfino, Marker Felt, Didot, Charcoal.

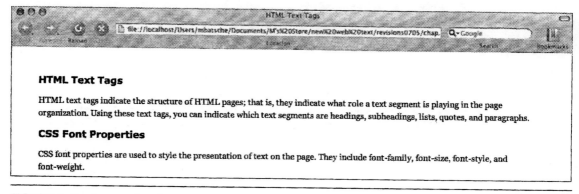

FIGURE 5–1 A page showing `font-family`

I've removed the centering for the heading in keeping with the recommendations on justification. Now let's add a font to the paragraph so that we can use something other than the default. The new style rule will look like this:

```
p{font-family:georgia,palatino,times,serif; line-
    height: 1.5;}
```

I'm going to save the external style sheet as `styles.css`. Now the head section on my HTML page will include a `<link/>` tag to link to my external style section. The head section now looks like this:

```
<!DOCTYPE html PUBLIC "-//W3C//DTD XHTML 1.0
    Transitional//EN" "http://www.w3.org/TR/
    xhtml1/DTD/xhtml1-transitional.dtd">
<html xmlns="http://www.w3.org/1999/xhtml">
<head>
    <title>HTML Text Tags</title>
    <link rel="stylesheet" href="styles.css"
    type="text/css" />
</head>
```

In a browser my new text style would look like Figure 5–1.

`font-size`

You can set the size of your font more precisely in CSS than with HTML, using a variety of options for the value of the `font-size` property:

❑ *Length designations:* points (`pt`), pixels (`px`), inches (`in`), centimeters (`cm`), or em's (`em`)

❑ *Keyword expressions* (defined by each browser): `xx-small`, `x-small`, `small`, `medium`, `large`, `x-large`, `xx-large`

SIDEBAR: THE MYSTERIOUS EM

You're probably familiar with measurements like points or pixels, but you may be wondering what the *em* measurement is. Ems are length units equal to the width of the letter *m* in whatever font you're using. Thus, the size of the em depends on the size of the font: `1em` would be the default font size; `2em` would be twice as big; and so on. Many Web designers prefer to use ems as font measurements because their size is roughly the same on all monitors, unlike points or pixels, which can vary widely from one computer to another. However, the em measurement has not been well implemented in various browsers and browser versions, and some browsers handle ems quite poorly. If you decide to use ems, be sure to check your pages in multiple browsers.

❏ *Relative expressions* (relative to the default font size): `smaller` or `larger`
❏ *A percentage* that indicates how much larger or smaller you want the font than its parent element

Both percentages and relative expressions refer to the size of the parent element, that is, the element in which this element is contained. Most text elements (i.e., paragraphs, headings, and so on) have preset default sizes in browsers (usually around 16px). You can use `smaller` or `larger` or percentages to indicate how you want your text to look in comparison to those preset sizes. For example, if you define `h1` as 80%, it will be 80 percent of the size that is set for the default `h1` in the browser. Although using these relative length designations will mean that type sizes are more similar in various browsers, type may still look different on different computer monitors because higher resolutions will show fonts in smaller sizes.

Other font-size measurements have other problems. Points measure type printed on a page, not displayed on a screen, and the implementation of point sizes differs from one browser to another. Although pixels are much more screen friendly, they cannot be enlarged via a screen zoom in Internet Explorer for Windows, which may cause problems for users who need larger type sizes. In many cases the most reliable font-size measurements may be the keyword expressions, which can be enlarged and which are relative to the default font size on the user's computer.

Let's create a new style rule on our external style sheet for a secondary heading. Since I set `h1` to `large`, let's try `medium` for `h2`.

```
h2{font-family: verdana, helvetica, sans-serif; font-
    size: medium;}
```

The rule would look like Figure 5–2 in a browser.

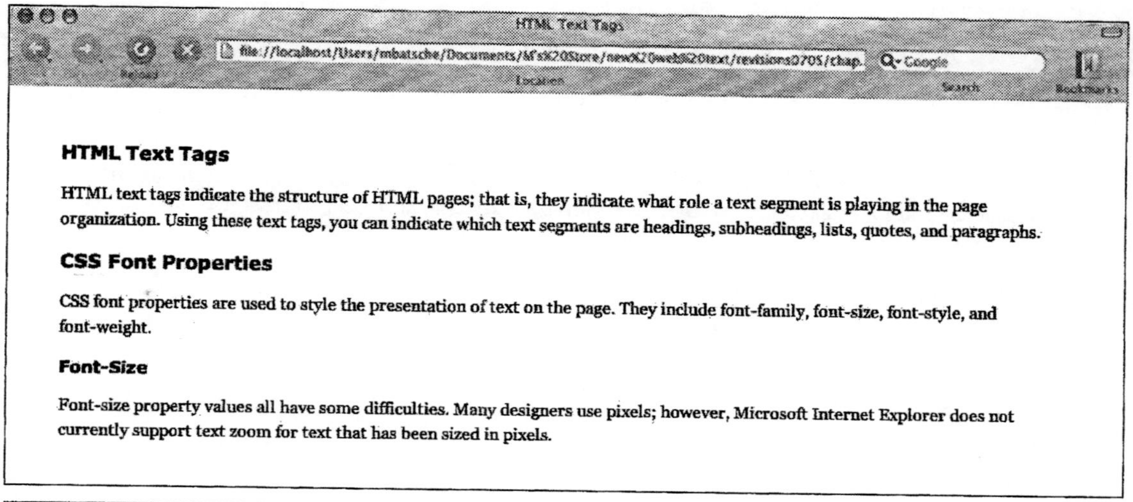

HTML Text Tags

HTML text tags indicate the structure of HTML pages; that is, they indicate what role a text segment is playing in the page organization. Using these text tags, you can indicate which text segments are headings, subheadings, lists, quotes, and paragraphs.

CSS Font Properties

CSS font properties are used to style the presentation of text on the page. They include font-family, font-size, font-style, and font-weight.

Font-Size

Font-size property values all have some difficulties. Many designers use pixels; however, Microsoft Internet Explorer does not currently support text zoom for text that has been sized in pixels.

FIGURE 5–2   A sample page showing font-size

font-style

To designate italic type, you can use the `font-style` property. Your choices for values are just two:

- ❑ `italic`
- ❑ `normal`, which removes all style designations

Remember, this is visual styling rather than adding italics to emphasize a word or indicate a source citation. For those applications you would use either `<em></em>` or `<cite></cite>`.

font-weight

CSS provides several levels of bold, using the `font-weight` property (although not all of these choices have been implemented in browsers yet). The choices for values are these:

- ❑ `bold`
- ❑ `bolder` or `lighter`, which sets the weight bolder or lighter than the default for the parent element
- ❑ Values from 100 to 900 in increments of 100 (Note: the number values have not been fully implemented yet in many browsers)
- ❑ `normal`, which removes all weight designations, including default bold-face for elements like headings

Let's create a third heading level for our external style sheet that uses these properties:

```
h3{font-family: verdana, helvetica, sans-serif; font-
   size: small; font-style:italic; font-weight:bold;}
```

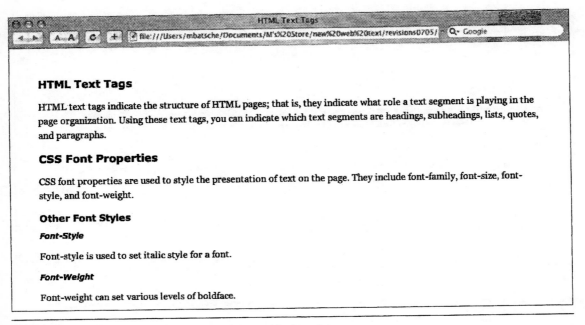

HTML Text Tags

HTML text tags indicate the structure of HTML pages; that is, they indicate what role a text segment is playing in the page organization. Using these text tags, you can indicate which text segments are headings, subheadings, lists, quotes, and paragraphs.

CSS Font Properties

CSS font properties are used to style the presentation of text on the page. They include font-family, font-size, font-style, and font-weight.

Other Font Styles

Font-Style

Font-style is used to set italic style for a font.

Font-Weight

Font-weight can set various levels of boldface.

FIGURE 5–3   A sample page with font-style and font-weight

In a browser it would look like Figure 5–3.

font

Using the `font` property allows you to set more than one font property at a time, instead of breaking the style rule up into `font-family`, `font-size`, and so on. However, you must list all of the values for these properties in a particular order, separated by spaces:

❑ Begin with the `font-style` and `font-weight` values (e.g., `italic`, `bold`). These values aren't required, but if you use them, they must come first. If you don't use them, you'll get the default values (i.e., `none`).

❑ After `font-style` and `font-weight` comes `font-size`, **which is required.** If you don't specify a font size, the style won't work.

EXERCISE: SET FONT PROPERTIES

Create some font styles for your Web pages now, using either an internal or an external style sheet. You can apply these font styles to most of the HTML text tags we covered in chapter 4: `p`, `h1`, `blockquote`, `ol`, `ul`, `dl`, `li`, `dt`, `dd`, `em`, `strong`, `cite`, `abbr`, and `acronym`.

❑ End with the `font-family` value, **which is also required**. Separate the font names with commas.

You can also include a `line-height` value within the `font` property if you want to. Put a forwardslash after the `font-size` value, and write the `line-height` value.

   If we use the `font` property rather than the individual properties, our CSS rule will look like this:

```
h3{font: italic bold small verdana, helvetica,
    sans-serif;}
```

CSS COLOR

CSS uses one main property to apply color to a wide variety of elements, including text. We'll cover the topic in more detail in chapter 7, where we discuss color, background color, and borders. You can use a variety of color models with CSS; again, we'll talk more about how they work in chapter 7, but these are your choices:

❑ Sixteen predefined colors: aqua, black, blue, fuchsia, gray, green, lime, maroon, navy, olive, purple, red, silver, teal, white, and yellow (Note: these colors will be interpreted differently in different browsers, so if you need a precise color, you'll want to use one of the other color systems)
❑ Hexadecimal color numbers (see chapter 7 for more information)
❑ RGB color designations (see chapter 7 for more information)

Let's add color to the `h1` style rule already on our external style sheet:

```
h1{font-family: verdana, helvetica, sans-serif; font-
    size: large; color:red;}
```

In a browser it would look like Figure 5–4.

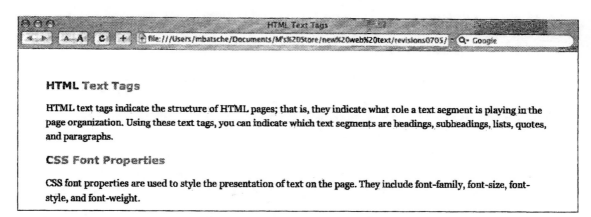

FIGURE 5–4 A page showing CSS color

CSS TEXT PROPERTIES

CSS divides the properties used with text into font properties and text properties. Font properties apply various styles to the fonts the words are written in; text properties apply various spacing options to the text blocks.

line-height

You've already been introduced to the `line-height` property in chapter 3; it's an important text property. CSS allows you to increase the space between the lines of your text so that you can avoid the dense, text-heavy look of the default single-space HTML pages. The property is `line-height` and as with `font-size`, you have some choices for values:

❑ *A length unit:* pixels (`px`), inches (`in`), centimeters (`cm`), points (`pt`), or ems (`em`)
❑ *A number that's multiplied by the font size to get spacing* (if you wanted double spacing, you'd use `2`, single spacing `1`, and so on; usually `1.5` to `2` is enough space)
❑ *A percentage that refers to the font size* (e.g., 100 percent would mean a space equal to the height of a line of text, or single space; 200 percent would mean double space; and so on)

You can define line height separately or as part of the `font` property. As a separate property, it looks like this:

```
p{font-family:georgia,palatino,times,serif; line-
    height:1.5; font-size:12px;}
```

To make `line-height` part of the `font` property, put a slash after the font size, and add the line height. It would look like this:

```
p{font: 12px/1.5 georgia,palatino,times,serif;}
```

text-align

As you saw in chapter 3 when we centered the heading, CSS includes options for aligning your text. The property is `text-align`, and your choices for values are these:

❑ `left` for left justification (the default justification)
❑ `right` for right justification
❑ `center` for centered text
❑ `justify` to align text on the left and right margins

Be careful about using `justify` on your Web pages. Spaces will be added between your words to make the lines of text stretch to the margins of your page, and the result may look strange, depending on how much text you have. Moreover, Web browsers lack the sophisticated hyphenation algorithms used in word processors and page layout programs, so they won't be able to break words between lines. For the most part, justified text is a poor choice for Web pages.

If we centered the h1 heading again, the CSS rule would look like this:

```
h1{font: large verdana, helvetica, sans-serif; text-
    align:center;}
```

text-indent

Web designers lamented the fact that the default HTML paragraph style didn't include paragraph indentation because many people find indented paragraphs easier to read and better looking than paragraphs separated by spacing. CSS allows you to create indented paragraphs using the text-indent property. You can use either of two measurements for values for your indentation:

❑ *A length measurement,* such as pixels (px), ems (em), inches (in), or centimeters (cm)
❑ *A percentage,* indenting the paragraph proportionally to the total paragraph width (e.g., text-indent:10% would set up a paragraph indent that was 10 percent of the paragraph width)

If we add indentation to the paragraph rule written earlier, the code would look like this:

```
p{font-family:georgia,palatino,times,serif; line-
    height:1.5; font-size:medium; text-indent:5%;}
```

In a browser the indented paragraphs would look like Figure 5–5.

HTML Text Tags

HTML text tags indicate the structure of HTML pages; that is, they indicate what role a text segment is playing in the page organization. Using these text tags, you can indicate which text segments are headings, subheadings, lists, quotes, and paragraphs.

CSS Properties for Text

CSS divides the properties that affect text into two categories: font and text. Font covers all the properties that apply to typography: type fonts, type sizes, type colors, boldface, and italic. Text covers all the properties that apply to spacing and alignment: leading (line space), tracking (word and letter spacing), paragraph indents, and alignment.

FIGURE 5–5   A page with text-indent

text-decoration

`text-decoration` is used chiefly for underlining links—or removing the underlining, which is the default style in most browsers. These are your choices for values in CSS:

- ❏ `underline` to underline text
- ❏ `overline` to place a line above text
- ❏ `line-through` to put a line through text, as if it had been crossed out
- ❏ `blink` to make text blink on and off (Note: most users find blinking text to be *very* distracting, and it can be a serious hazard for some users with epilepsy)
- ❏ `none` (the default)

We'll come back to the `text-decoration` property in chapter 8 when we discuss the CSS properties related to links.

letter-spacing and word-spacing

You can also use CSS to add or subtract space between letters or words. These properties can seriously affect the readability of your text, so they should be used with care. You might add `letter-spacing` for special effects in a page heading, for example, but you probably wouldn't add it to your running text. Your choices for values include these:

- ❏ *A length unit,* such as points (`pt`), picas (`pc`), pixels (`px`), inches (`in`), centimeters (`cm`), or ems (`em`)
- ❏ `normal` (the default)

If we add `letter-spacing` to the style rule for `h1`, it will look like this:

```
h1{font: bold large verdana, helvetica, sans-serif;
      text-align:center; color:red; letter-spacing:15px;}
```

In a browser it would look like Figure 5–6.

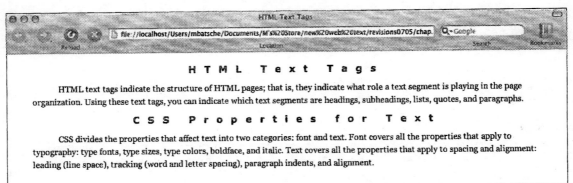

FIGURE 5–6 A page with `letter-spacing`

EXERCISE: SET TEXT PROPERTIES

Continue working on the style sheet for your text pages by setting text properties to go along with your font properties. You'll probably need to concentrate on more standard properties, such as `line-height`, `text-align`, and `text-indent`.

CSS LIST PROPERTIES

CSS provides you with some options for lists that weren't available with HTML. You can change list numbers and bullets, and you can use your own images for list markers.

list-style-type

The `list-style-type` property lets you choose a different number or bullet style rather than the default for your ordered and unordered lists.

- ❑ With an ordered list (use either `ol` or `li` for the selector), your choices are `decimal`, `lower-roman`, `upper-roman`, `lower-alpha`, `upper-alpha`, or `none` (`decimal` is the default). `lower-roman` and `upper-roman` will produce lowercase or uppercase roman numerals; `lower-alpha` and `upper-alpha` will produce lowercase or uppercase letters.
- ❑ With an unordered list (use either `ul` or `li` for the selector), your choices for values are `disc`, `circle`, `square`, or `none` (`disc` is the default). `disc` will produce solid bullets; `circle` will produce a circle outline.

A typical CSS rule would look like this:

```
ol li{list-style-type:upper-alpha;}
ul li{list-style-type:square;}
```

I'll add these styles to my style sheet, along with another style for the lists themselves:

```
ol, ul{font: medium/1.5 georgia,palatino,times,serif;}
```

If I apply this style sheet to the lists I wrote in chapter 4, the result will look like Figure 5–7.

list-style-image

You can use the `list-style-image` property to choose your own image for a bullet in an unordered list. But keep these custom bullets small—you don't

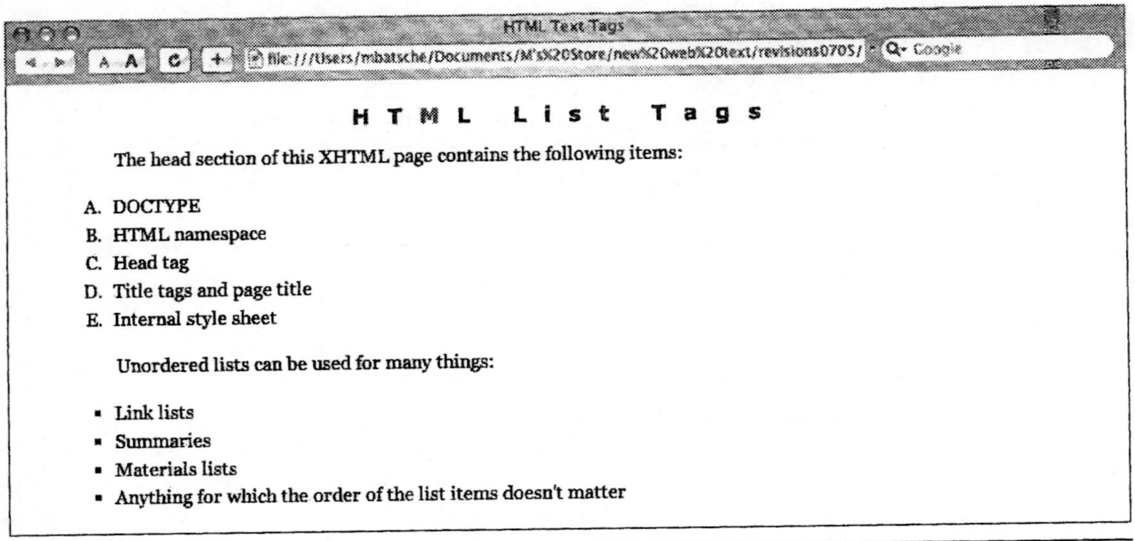

```
⊙⊙⊙                          HTML Text Tags
◄ ► │ A A │ C │ + │ file:///Users/mbatsche/Documents/M's%20Store/new%20web%20text/revisions0705/ │ Q▾ Google

                    H T M L    L i s t    T a g s

        The head section of this XHTML page contains the following items:

        A.  DOCTYPE
        B.  HTML namespace
        C.  Head tag
        D.  Title tags and page title
        E.  Internal style sheet

        Unordered lists can be used for many things:

        ▪  Link lists
        ▪  Summaries
        ▪  Materials lists
        ▪  Anything for which the order of the list items doesn't matter
```

FIGURE 5–7   A page using `list-style-type`

want to make the list look badly proportioned. Use the `ul` or `li` tags as your selectors here, with two choices for values:

- ❏ The URL of the image you want to use, expressed like this:
  `url(bullet.gif)` (use the name of your own custom bullet rather than `bullet.gif`)
- ❏ `none`, which creates a list with no marker at all

The CSS rule would look like this:

```
ul li{list-style-image:url(bullet.gif);}
```

If I add this rule to my style sheet, the unordered list would look like Figure 5–8.

EXERCISE: STYLE YOUR LISTS

In the previous chapter, I suggested that you identify lists within your content inventory. Now you can go back and add list styles to those lists, if those styles are appropriate for your content.

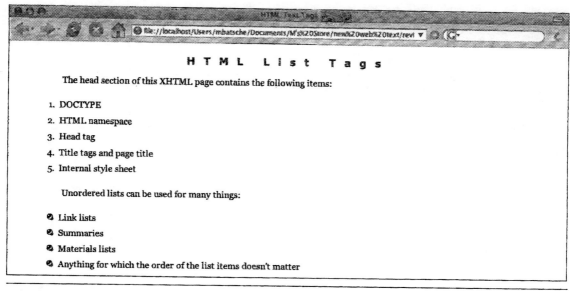

FIGURE 5–8 A page using `list-style-image`

CSS SPACING PROPERTIES

CSS allows you to add space to your pages without using the earlier design hacks that were standard with Web designers. All HTML elements can have two kinds of added space: margins and padding. Margins go around the outside of the element, whereas padding goes inside the edge of the element box. (For more about the CSS box model, see chapter 10). You've already used margins in chapter 2, and we'll talk more about padding in chapter 7.

Margins

Thanks to CSS, you can now include real margins on your pages, without having to use layout tables to set them up. You have several choices for values for your margins:

- ❑ *A length measurement:* pixels (`px`), centimeters (`cm`), ems (`em`), or inches (`in`)
- ❑ *A percentage* (margin width as a percent of the total page width)
- ❑ `auto`, which will set the margins to the default in the browser

You can place margins around each page, as we did in chapter 2, by using `body` as your selector. You can also set individual margins using these properties: `margin-left`, `margin-right`, `margin-top`, or `margin-bottom`. If I wanted to set only the left margin for the body text, for example, I could write my code like this:

```
body{margin-left:20px}
```

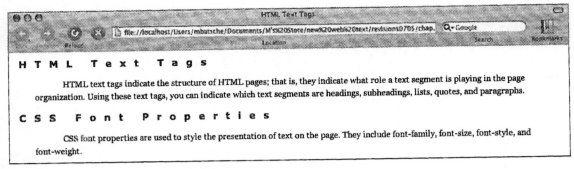

FIGURE 5–9   A page using margin-left

You can also set margins around individual elements on the page. For example, if you wanted your first-level headers to be placed closer to the left margin than your paragraphs, you could set the h1 margins separate from your paragraph margins. If I added those properties and values to my style sheet, the definitions would look like this:

```
p{margin-left:40px; font-family:georgia,palatino,
    times,serif; line-height:1.5; font-size:medium;
    text-indent:5%;}
h1{margin-left:10px; font:bold large verdana,
    helvetica, sans-serif; color:red; letter-
    spacing:15px;}
```

Figure 5–9 shows what that would look like in a browser (I've removed the margin I had placed around the entire page).

The margin property can use four different values if you want to set different margins on the page. The numbers go clockwise—top, right, bottom, and left—and the code looks like this:

```
body{margin: 20px 50px 0 70px;}
```

You can also use two values, which will set the top and bottom margins to the first value and the left and right margins to the second value. For example,

```
body{margin: 0 50px}
```

SIDEBAR: REMOVING THE EXTRA SPACE BETWEEN PARAGRAPHS

The default paragraph style in HTML puts extra space between paragraphs. However, if you want indented paragraphs, you won't want extra space in between them. You can use the margin property to remove the extra space. To do that, give your paragraphs a top and bottom margin of 0, like this:

```
p{margin-top:0; margin-bottom:0; text-indent:25px}
```

EXERCISE: SET YOUR MARGINS

Your margins should be consistent throughout your site. It will be easiest to make them consistent if you use an external style sheet that you apply to all your pages, or to all your pages that emphasize text. Try experimenting with your margins now until you find a combination of spacing that sets up your text effectively.

CLASSES AND IDS

As I explained in chapter 3, classes and ids allow you to create custom styles that you can apply to particular elements on a page rather than to all the elements. You can use a class to create a style you want to use more than once on a particular page. Let's say, for example, that you wanted all the paragraphs on your page to be indented, except for the first paragraph after a heading (indented paragraphs below headings can make text look unbalanced). You could create an intro class that you would apply to only the heading paragraphs. The style rule for your paragraphs could look like this:

```
p{font-family:georgia,palatino,times,serif; line-
     height:1.5; font-size:medium; text-indent:5%;}
```

The style rule for the intro class would have the same properties except for the indentation, like this:

```
.intro{font-family:georgia,palatino,times,serif; line-
     height:1.5; font-size:medium; text-indent:0;}
```

Here's the code for a page using the intro class:

```
<!DOCTYPE html PUBLIC "-//W3C//DTD XHTML 1.0
     Transitional//EN" "http://www.w3.org/TR/
     xhtml1/DTD/xhtml1-transitional.dtd">
<html xmlns="http://www.w3.org/1999/xhtml">
<head>
     <title>HTML Text Tags</title>
     <link rel="stylesheet" href="styles.css"
     type="text/css" />
</head>
<body>
<h1>CSS Properties for Text</h1>
<p class="intro">
CSS divides the properties that affect text into
two categories: font and text. Font covers all the
properties that apply to typography: type fonts,
```

```
type sizes, type colors, boldface, and italic. Text
covers all the properties that apply to spacing and
alignment: leading (line space), tracking (word and
letter spacing), paragraph indents, and alignment.
</p>
<p>
Classes and ids can also be used with text to
create custom styles that you can apply to
particular elements on a page rather than to
all the elements. Classes can be used to create a
style you want to use more than once on a particular
page. Styles created with ids can be used only once
on a particular page, which makes them best suited
for positioning.
</p>
<p>
You can use classes and ids with any HTML tag, as
long as the properties in the class or id work with
that tag. To use a class with an HTML tag, insert the
<em>class</em> attribute in the tag; to use an id
with an HTML tag, insert the <em>id</em> attribute in
the tag.
</p>
</body>
</html>
```

In a browser it would look like Figure 5–10 (I used the method described in the sidebar to remove the extra space between the paragraphs).

FIGURE 5–10 A page showing a CSS Class

You can use classes and ids with any HTML tag, as long as the properties in the class or id work with that tag. To use a class with an HTML tag, you insert the `class` attribute in the tag; to use an id with an HTML tag, you insert the `id` attribute in the tag.

You can apply the same class to several different tags on the same page if you wish. For example, if you developed a special class for emphasizing text, you could apply it to headings, list items, paragraphs, and so on. However, an id can be applied only once on a page.

Remember to keep your class and id names short, and don't use more than one word in the name (you can't have spaces in the name).

PROCESS FOR CREATING CSS PROPERTIES FOR TEXT

Use the following steps to create style rules for your text elements:

1. Choose a readable font in a size larger than nine pixels, use left justification, and limit the length of your text lines.

2. To create an external style sheet, open a text-only file, and write all of your style rules there. Place the following link tag in the head section of each page that uses these styles: `<link rel="stylesheet" href="styles.css" type="text/css" />` (insert the name of your style sheet file in place of `"styles.css"`).

3. To set the fonts for your HTML elements, use the `font-family` property, and list the fonts in order of preference, ending with a generic font type. You can also set other properties, including size (`font-size`), italic (`font-style`), and boldface (`font-weight`). Use the `font` property to set all font properties at once.

4. Set the color for your fonts using the `color` property and color names or hexadecimal numbers for values.

5. Use CSS text properties to set leading (`line-height`), alignment (`text-align`), and indentation (`text-indent`). More specialized graphic effects include underlining (`text-decoration`) and letter (`letter-spacing`) and word spacing (`word-spacing`).

6. Use list properties to change the numbers or bullets on your ordered or unordered lists. You can also substitute your own (small) graphic for the default list markers.

7. Set margins for the entire page using the `margin` property. Set margins for individual page sections or HTML elements using `margin-left`, `margin-right`, `margin-top`, and `margin-bottom`.

8. Use classes to set special styles that you use repeatedly on the page. Use ids to set special styles that you use only once on a page.

CHAPTER SUMMARY

CSS Basics

❏ Text should be designed to be legible, using left justification, a text font that is readable in a variety of sizes, a font size that can be enlarged in screen readers, and a text line length short enough for easy reading.

❏ CSS rules can be placed in the head section of an HTML/XHTML page, producing internal style sheets, or they can be placed on a separate page saved with a `.css` extension, producing external style sheets. They can also be used within an individual HTML tag with the `style` attribute.

❏ CSS rules define how HTML elements will look and behave. They consist of a selector to which the style is applied, a property defining what part of the style they affect, and a value for the property.

CSS Properties for Text

❏ CSS properties applying to text include font properties, text properties, list properties, and spacing properties.

❏ Font properties define font families, sizes, styles, weights, and colors.

❏ Text properties define line height (leading), alignment, indentation, various lining options, and letter and word spacing.

❏ List properties allow Web designers to select a different style for list markers.

❏ Spacing properties include margins, which create space outside the border of the element, and padding, which creates space inside the border of the element.

❏ CSS properties can be applied to classes and ids as well as to HTML tags.

CSS PROPERTIES IN THIS CHAPTER

PROPERTY	VALUE	EFFECT
font-family	Font name, `serif`, `sans-serif`, `cursive`, `fantasy`, `monospace`	Chooses a type font for a selector (cursive and fantasy are not recognized by some browsers)
font-size	Length unit: `xx-small`, `x-small`, `small`, `medium`, `large`, `x-large`, `xx-large`, `smaller`/`larger`, percentage	Sets the size for the type font
font-style	`italic`, `normal`	Italicizes type
font-weight	`bold`, `bolder`/`lighter`, number values (100–900), `normal`	Boldfaces type

Continued

Continued

PROPERTY	VALUE	EFFECT
`font`	`font-style, font-weight, font-size/line-height, font-family` (in that order)	Sets all font properties at once
`color`	Predefined color names, hexadecimal color numbers, RGB color formulas	Applies color to element
`line-height`	Length unit, number (i.e., 1 for single spacing, 1.5 for space and a half, 2 for double spacing, etc.), percentage	Sets space between lines of type
`text-align`	`left, right, center, justify`	Aligns text blocks
`text-indent`	Length unit, percentage	Sets paragraph indent
`text-decoration`	`underline, overline, line-through, blink, none`	Sets lining options for text blocks
`letter-spacing`	Length unit, `default`	Puts extra space between letters
`word-spacing`	Length unit, `default`	Puts extra space between words
`list-style-type`	`decimal, lower-roman, upper-roman, lower-alpha, upper-alpha, none` (ordered); `disc, circle, square, none` (unordered)	Changes the default list-item marker
`list-style-image`	`url(image.gif)` (substitute graphic file name for `image.gif`)	Allows use of custom list-item marker
`list-style`	`list-style-type, list-style-position, list-style-image`	Shorthand property that sets all list style elements at once
`margin` (also `margin-left, margin-right, margin-top, margin-bottom`)	Length unit, percentage, `auto`	Sets space outside the element box
`padding` (also `padding-left, padding-right, padding-top, padding-bottom`)	Length unit, percentage, `auto`	Sets space inside the element box

SOURCES

Clark, J. 2003. *Building accessible Websites.* Indianapolis, IN: New Riders.

Lynch, P. J., and S. Horton. 2001. *Web style guide: Basic design principles for creating Web sites,* 2nd ed. New Haven, CT: Yale University Press.

Nielsen, J. 2000. *Designing Web usability.* Indianapolis, IN: New Riders.

Zeldman, J. 2003. *Designing with Web standards.* Indianapolis, IN: New Riders.

Working with Images

WEB IMAGES

Images are integral parts of most Web sites. In this chapter we'll discuss finding images and using the correct image formats, along with the code for placing images and necessary attributes to be used with the img tag. We'll also discuss some possibilities for CSS and images, including floats and spacing.

FINDING WEB IMAGES

Images are available in a variety of formats from a variety of sources. Clip art is copyright-free art available either in print form or on CD-ROM. When you purchase a clip art collection, you purchase the right to use it in publications or on your Web pages. However, some clip art collections place restrictions on the number of images you can use within a page. Be sure to check the restrictions on any collection you use.

Not surprisingly, you can also find Web images on the Web itself. There are several clip art sites available, with clip art of varying quality. Free Web clip art is provided for Web page authors, usually by the artists, either to publicize their design services or to share their creations with others.

You can also download images from Web sites other than clip art sites, but you must be aware of the copyright issues involved (see the *Copyright and the Web* sidebar). If the Web site owner or designer is making these images available to you as a gesture of goodwill—that is, if the owner or designer expressly invites you to download the image—then feel free to download, although it's always wise to use a virus protection program to make sure the images aren't going to infect your computer. Several government sites (e.g., NASA) provide free images for downloads. And some cooperative photography sites, such as iStockphoto and stock.xchng, provide examples of photographers' work, which are provided either free or for low cost. In addition, images that are out of copyright, either because of age or nonrenewal of the copyright, are within the public domain and can be used without copyright violation. Many libraries have catalogs of public domain images, and you can find others using a search engine.

However, if the images on a Web site are there strictly for on-site use (i.e., to augment or illustrate the content of the site), then taking them without permission is equivalent to stealing them. If you proceed to use these images on your own site, you're violating the copyright of the original owner.

If you're using clip art from a clip art site, you should credit the source of your images on the page itself. Sometimes clip art sites will provide you with an icon to download and place on your pages to identify the source of your images, or you can link your site to the clip art site in order to allow other people to find images like yours.

One of the best ways to use original and effective images is to create your own. Either coupled with a scanner or used with an electronic clip art collection, a graphics program can increase your resources tremendously. The particular graphics program you use is less important than just getting used to a graphics program—once you become accustomed to one graphics program, others will seem less difficult.

SIDEBAR: COPYRIGHT AND THE WEB

It's easy to forget that information and images on the Web belong to other people, just like the information and images in a magazine. You can take images and information from the Web for your personal use, but you can't put those images and that information on your own Web pages without permission. Look at it this way: you can cut out a picture from a magazine and thumbtack it to your wall—that's personal use. But you can't cut out that picture and use it in your own magazine without paying the artist for its use—that would violate the artist's copyright. You would be using someone else's work for your own financial gain, and that doesn't constitute fair use. Here are some guidelines to keep in mind:

❏ Don't copy material from another site without permission, even if you give credit to the author.

❏ Never use someone else's picture, voice, performance, and so on without permission in writing.

❏ Don't assume that material posted on someone else's Web site belongs to the person who posted it—the author may be violating someone else's copyright without your knowledge.

❏ Be sure to use quotation marks if you quote a Web source directly, and always give full credit to your source even if you paraphrase someone else's words.

❏ Give credit to the sources of your information, both on the Web page itself and through a link to the original source.

❏ If you're mounting your pages on your school's Web server, be sure to check any campus policies on copyright and fair use.

USING IMAGES

The original specifications for HTML didn't include any tags for inserting images. In fact, images weren't added to the Web until the `img` tag was introduced in the Mosaic browser by Mark Andreeson in 1993. Now, however, it's hard to find an effective Web site without some kind of image use.

There are restrictions on using Web images that you should keep in mind. In particular, remember that Web browsers can read only a few types of image formats, and any image you include on your Web pages must be saved in one of those formats.

Web Image Formats

Before you can place an image on your Web page, it must be saved in the right format for the Web. It should also be saved in the format that's best for each particular image type. Currently, there are three formats supported by most Web browsers, with a fourth format under development for XML-based browsers.

GIF. GIF stands for Graphic Interchange Format, an image format developed by the Unisys Corporation. GIFs are highly compressible and can be squeezed down to small sizes. They also can mask one color, which means that you can create transparent backgrounds and place your image directly against the background of your page. (Actually, you're just covering up one color of the image, which means that you may not be able to eliminate all of the background color.) You can also use GIFs to create the equivalent of simple flip-book animations, putting several GIFs in a loop by using a GIF animation program, although few GIF animations are used on professional sites.

One drawback to GIFs is that they can use only 256 colors. Most photographs will be degraded in quality if you save them in GIF format because you won't be able to capture the subtle gradations in color from one part of the photograph to another (known as graduated tone). For this reason GIFs are used most frequently for line art: cartoons, drawings, sketches, and so on. Photographs are better saved as JPEGs unless the photograph is a small one in which the loss of detail won't matter and the small file size is a major consideration.

JPEG. JPEGs use a format developed by the Joint Photographic Experts Group especially for photographs on the Web. They're designed to handle the graduated colors of photographs and to produce relatively small files. JPEGs deal with the brightness and color of photographs separately, compressing the more subtle gradations in color. It's a *lossy* process, meaning JPEGs lose information in compression. Thus, JPEG images won't have the same quality as the original image, although their quality may be enough for your Web site. You can save your JPEGs at various levels of quality (usually low, medium, high, and maximum). The better the quality, the sharper the picture—and the larger the file size. In general, with all Web images you'll try for the lowest quality setting you can get by with, because you'll want to keep your file size as small as you reasonably can to limit the download time.

JPEGs don't handle flat colors well; the process is designed to work with graduated color. Thus, for images with flat color, like drawings, GIFs or PNGs are better choices since they handle flat color better and are more compressible.

PNG. PNG, which stands for Portable Network Graphics, is the third image standard supported by the W3C, although it's not well supported by all browsers, particularly versions of browsers prior to 2000. PNG was developed as a royalty-free alternative to the GIF format (software producers must pay royalties to Unisys to include GIF images in their programs), and it has some advantages as an image format as well. First, PNG images can have 24-bit color depth, although these image files may be quite large. However, these images lose virtually no quality, preserving all their detail and color values. PNG images that use 8-bit color depth use indexed color as GIFs do, but they can be compressed up to 25 percent more than a GIF file of the same image. PNG also provides greater control of transparency, allowing some images to eliminate backgrounds more effectively than GIFs do, for example. Furthermore, saving, restoring, and resaving a PNG image will not degrade its quality. However, the PNG format does not support animation, as GIF does.

SVG. This additional Web image format exists but is not widely used as yet. SVG stands for Scalable Vector Graphics; it is a vector image format that allows two-dimensional images to be displayed on pages written with XML. Like most vector images (and in contrast to GIF, JPEG, and PNG images, which are raster—that is, bitmapped—images, which always remain a specified size), SVG images can be scaled in size and resolution to fit the browser window. In addition, SVG images have smaller file sizes than GIFs or JPEGs. However, browser support for SVG images is limited; they are currently viewable only on browsers using a plug-in from Adobe.

Which Image Format to Choose

GIFs and PNGs handle flat color more effectively than they do graduated color. Although 24-bit PNG images could conceivably handle the graduated color of a photograph, the resulting file size would be considerable. The choice of which image format to use on your page is actually simple:

IF . . .	THEN . . .
Your image is a photograph	Use JPEG
Your image is something other than a photograph	Use GIF or PNG

USING IMAGES ON WEB PAGES

You've already been introduced to the `img` tag for inserting images (see chapter 2), but there's one point that should be stressed: these images aren't really placed on your pages. The `img` tag is actually a direction that tells the browser to request a particular image file from the server. Once the server supplies the image file, the browser displays it on the page in the place where the tag appears.

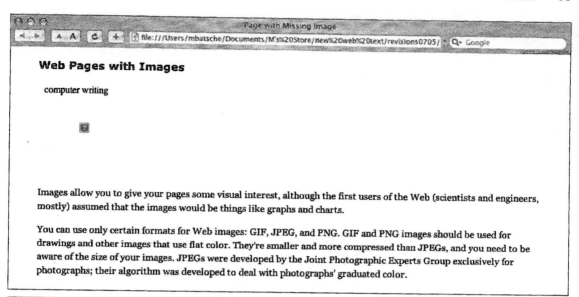

FIGURE 6–1   Missing image icon

It's easy for beginning Web designers to forget that the image isn't actually embedded in the page, particularly if you're using an authoring program that shows you the image as if it were part of the page. To make that image appear at that point on the page, you need to supply the image file along with the HTML file for the page. When you upload the files to a Web server, the image file has to be uploaded along with the page file. Otherwise, when the browser opens your page, your reader may see a missing image icon, as shown in Figure 6–1.

SIDEBAR: FILE SIZE ON THE WEB

The issue of file size is important, particularly with images. Although many people now use high-speed Internet connections like cable modems and DSL, many still use dial-up modems. According to the U.S. Department of Commerce, in 2003 high-speed Internet hookups accounted for only around 20 percent of all Internet connections in the U.S. That's a good reason to keep your file sizes relatively small, no larger than 60K or 70K per page as a rule (and some designers suggest 30K). A user who's interested may be willing to wait for thirty seconds or longer while your page downloads. However, some studies have indicated that ten seconds is the longest that many users will wait for some kind of action. You can test yourself to see how this works: How fast do your favorite Web pages download on your Internet connection? How long are you willing to wait for a page to be completely visible before you feel like moving on?

Converting Images

If you've scanned in an image or chosen an image from a clip art CD, it may use TIFF or EPS or another image format. Your first step should be to determine the current format of your image file. You can look at the image file on your desktop to see what the extension is, or you can open the image in your graphics program and check it there. Different graphics programs use different processes to convert images from one format to another. Consult the documentation on your graphics program to find out how to do it.

Working with a JPEG

When you choose JPEG as the format in which you want to save a photograph, you'll probably be given a choice of quality levels (usually low, medium, high, and maximum). If you want to experiment to see how low you can make the quality level and still get an acceptable image, try saving your image at various levels under different names (e.g., "imagelow," "imagemed," "imagehigh," "imagemax"), and then look at the different versions side by side. Remember, you want an image that doesn't take up much memory, but not one that's hard to recognize. Many browsers now also support the Optimized setting for JPEG files; these files will be slightly more compressed than standard JPEGs. You can also use Progressive JPEGs, which download in segments, a good format for large photographs.

Working with a GIF

The GIF image format uses a color mode called indexed color. The color index in this case contains a palette of 256 colors, reducing the file size of the image by limiting the number of colors that it can use. In some graphics programs you can choose the number of colors to be used in the image and thus reduce the image size still further. For example, if the original image uses 256 colors, you can try reducing it to 128 or fewer (e.g., 64 or 32) to see if the quality remains high enough to be used on your site. As with JPEGs, you'll try to balance the smallest file size with the level of quality you need. Some graphics programs also allow you to choose a lossy option for GIFs, which will further compress your file. According to Dave Shea and Molly Holzschlag (2005), "A very small amount of lossiness, 5 or 6 percent, can often drop an extra 20 or 30 percent off a Gif's file size without noticeably affecting quality" (129).

Saving an Image

Once you've converted your image to GIF or JPEG or PNG, you must save it with an extension that indicates what type of image it is. The extensions for images are .gif, .jpg, and .png. The complete image name would look like this: image.gif, image.jpg, image.png. Incidentally, the rules for naming and saving your files also apply to image files—keep the file name brief and don't use any spaces. These extensions tell the browser what type of information the file contains so that the browser knows how to interpret the file.

Resolution refers to the sharpness of an image: a high-resolution image will be very sharp, whereas a low-resolution image may be slightly blurry. When you scan an image for print, you usually scan at a relatively high resolution, frequently 200 dpi (dots per inch) to 300 dpi. However, the resolution of modern computer monitors varies from 40 dpi (a 21-inch screen set to 640x480 pixel resolution) to 165 dpi (a 14-inch screen set to 1600x200 pixel resolution). Many Web designers suggest saving your images at 72 dpi as a generic number that will display an image on most monitors in a size that approximates the actual size of the image in inches. This generic figure is based on the assumption that you want to maintain the same size as the original. If you actually want a different size, scanning guides both in print and on the Web can provide more information about image resolution for the computer monitor.

There's one important point to keep in mind, however: *simply attaching the extension to the image name will not convert the image to another mode.* In other words, if you haven't converted your image to a GIF in your graphics program, simply adding `.gif` to the image name won't do it.

You must also use the right extension for the format of your image. If your image is a JPEG and you forget and name it with `.gif,` not only will it not be converted to a GIF, but it won't show up in the browser! Most graphics programs can automatically add the extension when you save an image. Check the documentation on your own graphics program to find out how this feature is enabled.

PLACING AN IMAGE ON A WEB PAGE

As you saw in chapter 2, the basic tag for inserting an image on your page is `img`, which stands for *image*. The `img` tag is always used with an attribute, `src`, which stands for *source*. The value that goes with `src` is the name of the image file. Thus, the whole tag looks like this in XHTML (since `img` has no close tag):

```
<img src="image.gif" />
```

As I indicated in chapter 2, the `img` tag has several attributes beyond `src,` some of which are required for well-formed XHTML.

Adding Alternate Text

For visually disabled visitors using screen readers, who will not be able to see your images, HTML provides an alternative. The `alt` attribute is used to provide a brief text description of the image that will be read by the screen reader. It will also provide information for users who don't have images-enabled browsers, as is the case with many Web-enabled devices like cell phones. In addition, `alt`

SIDEBAR: THE OBJECT TAG

XHTML 2.0 proposes to deprecate the img tag in favor of the object tag. The object tag has been part of the HTML specifications for some time but has not been widely supported by browsers. Rather than the src attribute, the object tag uses a data attribute, which specifies the file name and location, as in the following: <object data="image.gif">. The object tag can also be used with various types of multimedia files and Java applets (i.e., small programs), which makes it more versatile and multipurpose than the img tag. The major stumbling block at this point is browser support, which is currently sketchy at best. Still, the object tag may come to be the standard in the future when browser support is more universal. We'll discuss the object tag in more detail in chapter 12, "Using Multimedia."

attributes are helpful for users with slow connections; the value of the alt attribute is visible as the image downloads and provides the user with some idea of what the image will be. And if for some reason your image doesn't show up, the alt attribute will tell the user what's supposed to be there. If you look back at Figure 6–1, you can see the alt attribute value, "computer writing," above the missing image icon. The alt attribute appears within the image tag, like this:

```
<img src="compwriter.gif" alt="computer writing" />
```

SIDEBAR: TIPS FOR ALT ATTRIBUTES

1. Provide an alt attribute for every image; alt attributes are required for XHTML.
2. If your image doesn't convey any information (e.g., a square of color), you can add an empty value for the alt attribute, like this: alt= "". However, don't use the empty alt for images that convey information; instead, supply the text description.
3. If you use an image for a link, you must provide an alt attribute that indicates the link destination.
4. Keep the alt text brief (e.g., "computer writing" rather than "picture of a computer with arms writing on a notepad"). Although there's no set limit on the size of the alt text, accessibility expert Joe Clark (2002) suggests using a conventional limit of 1,024 characters (1K) or less, since longer alts may not be fully displayed if images are turned off (63).
5. You don't need to indicate that a picture is a link; screen readers will identify links automatically (e.g., put "Communication Department home page" rather than "link to Communication Department home page").
6. Don't use the image file name as the alt text; briefly sum up the image content or its function.

Your HTML should look something like this:

```
<!DOCTYPE html PUBLIC "-//W3C//DTD XHTML 1.0
    Transitional//EN" "http://www.w3.org/TR/xhtml1/
DTD/xhtml1-transitional.dtd">
<html xmlns="http://www.w3.org/1999/xhtml">
<head>
    <title>Page with Images</title>
    <link rel="stylesheet" type="text/css"
    href="styles.css" />
</head>
<body>
    <h1>Web Pages with Images</h1>
    <img src="compwriter.gif" alt="computer
    writing" />
    <p>Images allow you to give your pages some
visual interest, although the first users of the
Web (scientists and engineers, mostly) assumed
that the images would be things like graphs and
charts.</p>
    <p>You can use only certain formats for Web
images: GIF, JPEG, and PNG. GIF and PNG images should
be used for drawings and other images that use flat
color. They're smaller and more compressed than JPEGs,
and you need to be aware of the size of your images.
JPEGs were developed by the Joint Photographics
Experts Group exclusively for photographs; their
algorithm was developed to deal with photographs'
graduated color.</p>
</body>
</html>
```

The page would look like Figure 6–2 in a browser.

Height and Width for Images

It's a good idea to also add `height` and `width` attributes to your image tags, providing the image's dimensions in pixels. If the browser is given the dimensions of the image, it can allot space for it, and the user will see the page layout more quickly. Text blocks will be placed around the space allotted for the image, and the images themselves will be "flowed in" after they finish downloading. You can find the dimensions of your image in a graphics program like Photoshop, or an authoring program can fill them in automatically. The code looks like this:

```
<img src="compwriter.gif" width="136" height="144"
    alt="computer writing" />
```

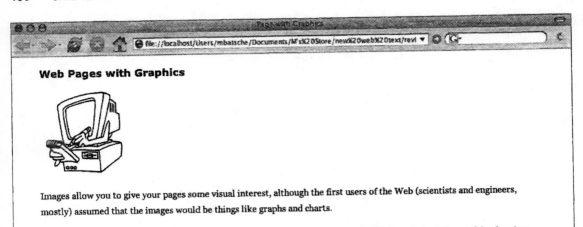

FIGURE 6–2   A page with an image

EXERCISE: ADD IMAGES

Try adding some images to your Web page(s). Be sure to save them in a Web format and place them in the same folder as your HTML files. Then write the img tag into your code. Use alt attributes for all your information-bearing images, and include the height and width attributes for each.

Floating an Image

As you can see in Figure 6–2, the default placement for images can leave some unsightly gaps, with the text ending above the image and beginning again below it. To get rid of these gaps, you can use CSS to wrap text around an image. The property here is float, and the values are left and right.

The float property removes an element from the normal flow of elements on your page and formats it as a block-level element. It you set the float property of an image to left, the image will be moved left until it reaches the margin, padding, or border of another block-level element, usually the body of your page or a page section like a div. The rest of your content will wrap around on the right side. (We'll discuss the float property in greater detail, along with more information about document flow and the CSS box model, in chapter 10.)

If you want your picture on the left and your text on the right, you could write the tag like this:

```
<img src="image.gif" alt="image" style="float:left;">
```

However, if you want all of your images on the left, you could use img as the selector for a style rule like this:

```
img{float:left;}
```

You could also float all the images within a particular section of your page by using a descendant selector, like this code, which floats all the images within an element that has the id "intro":

```
#intro img{float:left;}
```

Here's the XHTML code with the float added:

```
<!DOCTYPE html PUBLIC "-//W3C//DTD XHTML 1.0
    Transitional//EN" "http://www.w3.org/TR/xhtml1/
    DTD/xhtml1-transitional.dtd">
<html xmlns="http://www.w3.org/1999/xhtml">
<head>
    <title>Page with Images</title>
    <link rel="stylesheet" type="text/css"
    href="styles.css" />
</head>
<body>
    <h1>Web Pages with Images</h1>
    <img src="compwriter.gif" width="136"
height="144" alt="computer writing" style= "float:
left" />
    <p>Images allow you to give your pages some
visual interest, although the first users of the Web
(scientists and engineers, mostly) assumed that the
images would be things like graphs and charts.</p>
    <p>You can use only certain formats for Web
images: GIF, JPEG, and PNG. GIF and PNG images
should be used for drawings and other images that
use flat color. They're smaller and more compressed
than JPEGs, and you need to be aware of the size of
your images. JPEGs were developed by the Joint
Photographics Experts Group exclusively for photo-
graphs; their algorithm was developed to deal with
photographs' graduated color.</p>
</body>
</html>
```

Figure 6–3 shows what it would look like in a browser.

In this case I added the float by using a style attribute within the img tag. I could also add it to the style sheet, using img as the selector (if I wanted all of

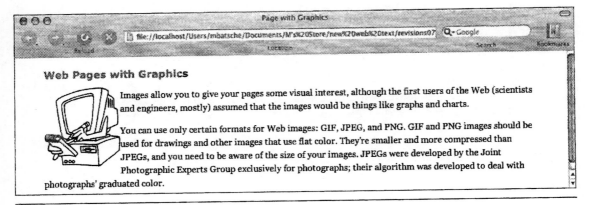

FIGURE 6–3   A page with a floated image

my images to float on the left) or creating a special class that I could associate with left-floating images.

Clearing a Float

Unlike other HTML elements, floated elements can stick out of the bottom of their containing blocks if the blocks are too small to contain them completely. Figure 6–4 provides an example, and again, you can refer to the information about the CSS box model in chapter 10. Here the image in the first div extends into the div below it, pushing the second image and text to the side. In many cases that's a good idea; you may want text to continue wrapping around the image in adjacent paragraphs, rather than having more unsightly spaces around the image.

However, in cases where you don't want the image to interfere with the content that follows it, you'll need to clear the float. Eric Meyer (2003) outlines several

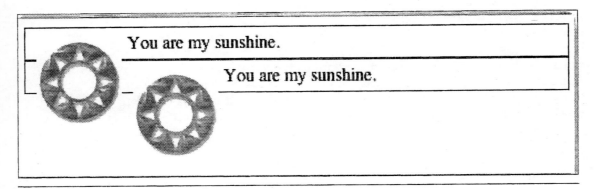

FIGURE 6–4   An uncontained float

methods for clearing a float; see http://www.complexspiral.com/publications/ containing-floats. One of these methods uses the `clear` property, which CSS provides. However, you'll need to create a block element with which to apply the `clear` and then insert that element into the block that contains the image. In effect, the `clear` property stretches the containing block so that it completely contains the floated image. `clear` has three possible values: `left`, `right`, and `both`.

The code used to produce Figure 6–4 follows:

```
<!DOCTYPE html PUBLIC "-//W3C//DTD XHTML 1.0
    Transitional//EN" "http://www.w3.org/TR/xhtml1/
    DTD/xhtml1-transitional.dtd">
<html xmlns="http://www.w3.org/1999/xhtml">
<head>
    <meta http-equiv="content-type" content=
    "text/html; charset=utf-8" />
    <title>Extending Float</title>
    <style type="text/css">
div.item{border: 1px solid; padding:5px}
div.item img {float:left; margin:5px}
</style>
</head>
<body>
<div class="item">
<img src="sun.gif" width="77" height="79" />
You are my sunshine.
</div>
<div class="item">
<img src="sun.gif" width="77" height="79" />
You are my sunshine.
</div>
</body>
</html>
```

To clear the float, I'll create a new "clearer" class: `div.clearer {clear:left;}`. I'll add it to the HTML, using a nonbreaking space—` `—to give the `clearer` div some invisible content:

```
<!DOCTYPE html PUBLIC "-//W3C//DTD XHTML 1.0
    Transitional//EN" "http://www.w3.org/TR/xhtml1/
    DTD/xhtml1-transitional.dtd">
<html xmlns="http://www.w3.org/1999/xhtml">
<head>
    <meta http-equiv="content-type" content=
    "text/html; charset=utf-8" />
    <title>Extending Float</title>
```

```
        <style type="text/css">
div.item{border: 1px solid; padding:5px;}
div.item img {float:left; margin:5px;}
div.clearer {clear:left;}
</style>
</head>
<body>
<div class="item">
<img src="sun.gif" width="77" height="79" />
You are my sunshine.
<div class="clearer"> </div>
</div>
<div class="item">
<img src="sun.gif" width="77" height="79" />
You are my sunshine.
<div class="clearer"> </div>
</div>
</body>
</html>
```

In a browser it would look like Figure 6–5. You can also use the float property with text, floating columns beside each other. We'll talk more about using floats in page layout in chapter 10.

Putting Space around an Image

Once you float your images, you may feel that the text and the image are a little too close together, particularly if you have a small image and a lot of text. Notice that in Figure 6–3 the text is almost touching the image. You can add some space

You are my sunshine.

You are my sunshine.

FIGURE 6–5   A contained float

EXERCISE: FLOAT IMAGES

You already have some images inserted on your pages; try styling them to float either left or right. If all your images float on the same side, you can add img to your style sheet. However, if they vary between right and left, use the style attribute. If necessary, clear any floats that interfere with each other.

around your images by using margins or padding in your style sheet, much as you did with your text in chapter 5. Just use img as the selector and put some space around it, like this:

```
img{margin: 10px;} or img{padding:10px;}
```

Your external style sheet will look something like this:

```
body{margin: 20px 50px 20px 50px;}
h1{font: bold large verdana, Helvetica, sans-serif;
    color:red; text-align:left;}
p{font-family:Georgia,palatino,times,serif; line-
    height:1.5; font-size:medium;}
img{margin:10px;}
```

And in the browser it would look like Figure 6–6.

EXERCISE: ADD SPACE

Go back to the pages where you floated images. Add img to your style sheet, and set a margin or padding for 10px or so.

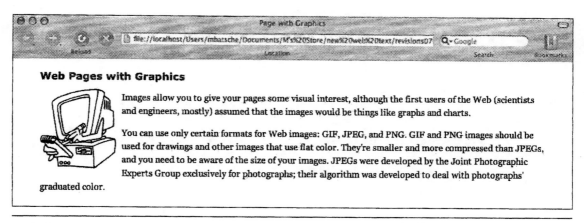

FIGURE 6–6 A page with an image with spacing

OTHER IMAGES: IMAGE MAPS AND SLICED IMAGES

Both image maps and sliced images were common design features in the late nineties. Image maps are large images with active areas that serve as links. They're frequently created in Web graphics programs like ImageReady and Fireworks. In these programs a designer draws shapes on the parts of the image that are to be active links and then fills in the blanks in a preformatted code section. The code contains the coordinates of the active shapes and the URLs to which the areas are linked and is placed between `<map></map>` tags. The `img` tag for the image must include the `usemap` attribute with a value that is the map name.

Image maps are usually large images and are thus potentially slow loading. From an accessibility standpoint, the problem with image maps is that they work most effectively in image-based browsers. Joe Clark (2002) maintains that `alt` and `title` attributes should be added to "any area that is graphical in nature" (189), including live areas on image maps. However, Jeffrey Zeldman (2003) points out that image maps "conveyed no structural meaning. They were simply pictures you could click on if you had the right equipment (including physical human equipment)" (191). For these and other reasons, the use of image maps has declined over the past few years. Although there are still some limited uses for image maps (including actual maps with clickable locations), think carefully before committing yourself to this format.

"Slicing" was another way of dealing with large images on older, usually table-based Web sites. The image was cut apart in a Web images program and then reassembled on HTML tables. Recurring elements could be saved in the browser's memory cache, and the entire image could be made accessible by adding an `alt` tag. Joe Clark (2002) recommends using empty `alt`s for most of the slices, choosing one slice to serve as representative of the whole, and adding an `alt` text to that single slice (113). As you might guess, if these sliced images included text, trying to keep them up-to-date was a fairly nightmarish chore. Moreover, search engines cannot read images, so using images alone to present text is rarely a good idea. In addition, using images that have been sliced into a large number of smaller pieces and then placed on complex tables automatically adds to the amount of code on the page, increasing the page size and thus the download time. Rather than creating complicated sliced designs, try using a simpler CSS layout. Most users will appreciate the savings in bandwidth requirements.

PROCESS FOR INSERTING AND STYLING IMAGES

Use the following steps to place images on your pages:

1. Find or create images in the correct file format: JPEG for photographs and GIF or PNG for line art.
2. Upload your image files along with your HTML files.

3. Place the image on the page using the `<img src="image.gif">` tag and attribute.
4. Add `alt` attributes to all of your information-bearing images.
5. Add `height` and `width` attributes to all images, with the values in pixels.
6. Use the `float` property, either on your style sheet (with `img` as the selector) or with the `style` attribute, to wrap the text around the image.
7. If the float interferes with elements that follow the image on the page, clear the float.
8. Put space around your images by using `img` as the selector on your style sheet and `margin` or `padding` as the property.

CHAPTER SUMMARY

Image Basics

❑ Web images are available from a variety of sources, including clip art collections and collections of public domain images provided on the Web. However, designers should obtain permission if they want to use graphics taken from a Web site that does not provide the images for their use.

❑ Images used on the Web must be in one of three formats: GIF, JPEG, or PNG. JPEGs are used for photographs; GIFs or PNGs are used for everything else.

❑ Images are not really placed on a Web page; the browser is given the name of the image file to load at a particular point on the page. You must supply image files along with HTML files when you upload a Web site.

❑ Images that are scanned in or taken from clip art collections must be converted to a Web format before they can be used.

Image HTML

❑ The basic image tag is `img`, which stands for image. It is always used with the `src` attribute, which provides the source for the image, that is, the file name and location.

❑ You should always provide an `alt` attribute for your images, giving a text version for browsers that cannot view images.

❑ You should add `height` and `width` attributes to your `img` tag. If the browser has the dimensions of the image, text can be loaded and read while the image is downloading.

Image CSS

❑ You can use the CSS `float` property to place an image alongside text rather than on separate lines. However, you may need to use the `clear` property to keep the floated image within its containing block.

❑ To put space around your images, make `img` a selector on your style sheet, and add either a margin or padding.

HTML/XHTML TAGS IN THIS CHAPTER

TAG	EFFECT
`<img>` (HTML) `<img />` (XHTML)	Places images on a page; must be used with `src` attribute
`<object> </object>`	Places images on a page; must be used with `data` attribute (replaces img in XHTML 2.0).

HTML/XHTML ATTRIBUTES IN THIS CHAPTER

ATTRIBUTE	VALUE	EFFECT
`src`	URL of image source	Gives browser location and file name of image; used with `<img>` (HTML) or `<img />` (XHTML)
`data`	URL of image source	Gives browser location and file name of image; used with `<object> </object>`
`alt`	Text description of image	Provides text equivalent of image for browsers that cannot use images
`height`	Height of image in pixels	Provides browser with image dimension
`width`	Width of image in pixels	Provides browser with image dimension

CSS PROPERTIES IN THIS CHAPTER

PROPERTY	VALUE	EFFECT
`float`	`left, right`	Floats an image alongside text
`clear`	`left, right, both`	Clears the float so that it doesn't influence other elements on the page

SOURCES

Clark, J. 2002. *Building accessible Websites*. Indianapolis, IN: New Riders.

Meyer, E. 2003. Containing floats. http://www.complexspiral.com/publications/containing-floats/

Shea, D., and M. Holzschlag. 2005. *The Zen of CSS design*. Indianapolis, IN: New Riders.

Zeldman, J. 2003. *Designing with Web standards*. Indianapolis, IN: New Riders.

Working with Colors and Backgrounds

USING COLOR

Thanks to CSS there are a variety of effects available from using colors for both foreground and background, as well as new possibilities for background images and borders. In this chapter we'll discuss the Web color models and ways to use color effectively. We'll also discuss the effective use of background images and borders.

COLORS ON THE WEB

There are three ways of indicating the colors you want to use on your Web site. First there are sixteen predefined color names: aqua, black, blue, fuchsia, gray, green, lime, maroon, navy, olive, purple, red, silver, teal, white, and yellow. All browsers will recognize and display the colors associated with these color names if you use them as the value for the `color` property. However, different browsers will use slightly different shades of these colors. Thus, red in Opera may be slightly different from red in Netscape. In HTML 4.01 and XHTML, the W3C recommends a different approach, using hexadecimal numbers rather than names to represent colors, so it's a good idea to become familiar with that system.

Hexadecimal Colors

Color is related to the bit depth of computer monitors. Older computer monitors had 8-bit displays; that is, 8 bits of memory were dedicated to each pixel in the computer monitor. This meant that the monitor could display a maximum of only 256 colors. Netscape developed a color system that limited the number of colors displayed on the Web to 216 (the original 256 minus 40 colors that the Windows operating system reserved for its interface). This color system became known as the browser-safe palette.

The browser-safe palette represents colors as numbers that refer to the percentages of red, blue, and green that make up the color you specify. To use a hexadecimal number, you begin with a pound sign (#) followed by the combination of letters and numbers that make up the color designation. For example, in the hexadecimal number #cc6600, cc refers to the shade of red, 66 refers to a shade of green, and 00 refers to a shade of blue. Added together, these three values produce a golden brown shade. Although you can find conversion tables for hexadecimal numbers, the easiest way to use hexadecimals is to look at a hexadecimal color chart, choose the shade you want, and then insert that number into your style sheet. Most authoring programs include hexadecimal tables that you can consult, and hundreds of them are available on the Web. There are also print versions available, but since the print versions use the CMYK color model rather than RGB, the color you end up with may be slightly different from the color you see on the printed page.

Although the complete hexadecimal number is six digits long (e.g., cc6600 or ff3366 or 00ff00), the W3C has developed a shorter notation for hexadecimal numbers, which uses only three digits, one from each pair (e.g., c60 or f36 or 0f0). You can use this form on your style sheets; the style rule looks like this:

```
p{color:#c60;}
```

Remember to include the pound sign before the number, or the color may not come up in some browsers. The code on an external style sheet would look like this:

```
body{margin: 20px 50px 0 50px; background-color:#ff6;}
p{font-family: verdana, helvetica, trebuchet, sans-
    serif; font-size:medium; color:#30f; background-
    color:#ccc;}
h1{font-family: verdana, helvetica, trebuchet, sans-
    serif; font-size:large; color:#30f; background-
    color:#f6f;}
em{font-family: verdana, helvetica, trebuchet, sans-
    serif; font-size:large; color:yellow; background-
    color:#c06;}
h2{color:#600; font-family: verdana, helvetica, tre-
    buchet, sans-serif; font-size: large;}
```

RGB Color Values

For CSS2 and CSS3, the W3C has developed a new system of color notation that provides a wider range of colors than the older browser-safe palette. Modern computers have 16- and 32-bit color depth, allowing them to display thousands or millions of colors. The newer color notation system relies on RGB values, producing a wider range of color shades. The format is rgb(nnn,nnn,nnn) in which "nnn" stands for a three-digit value for red, green, and blue.

In this new system of color notation, the hexadecimal color just described (#cc6600) would be written rgb(204,102,0). Once again, it's easier to take these values from a color chart or from a graphics program that specifies RGB values for a color. The code would look like this:

```
p{color:rgb(204,102,0);}
```

The code on an external style sheet would look like this:

```
body{margin: 20px 50px 0 50px; background-
    color:rgb(255,255,102);}
p{font-family: verdana, helvetica, trebuchet, sans-
    serif; font-size:medium; color:rgb(51,0,255);
    background-color: rgb(204,204,204);}
h1{font-family: verdana, helvetica, trebuchet, sans-
    serif; font-size:large; color:rgb(51,0,255);
    background-color:rgb(255,102,255);}
em{font-family: verdana, helvetica, trebuchet, sans-
    serif; font-size:large; color:yellow; background-
    color:rgb (204,0,102);}
h2{color:rgb(102,0,0); font-family:verdana, helvetica,
    trebuchet, sans-serif; font-size: large;}
```

This new color notation system provides designers with a much broader range of colors than ever before, up to 1068 color codes. However, these colors may not come up on all computers; a computer monitor that is set to display only 8-bit color will not be able to display some of them. Older browsers may not be able to recognize this particular color notation system. And if the browser does recognize the color system but the monitor is set to display 8-bit color, the browser will "dither" the colors. That is, it will place pixels of two colors close to one another on the screen to create the illusion that a third color is present (e.g., pixels of black and white can be placed close together to represent gray). Dithering reduces the sharpness of an image and may make the color seem grainy, but the color may still be close enough for your purposes. The color notation system you use should depend on the type of browser and monitor you expect most of your visitors to be using.

BACKGROUND COLORS

The CSS method of setting background colors provides much more flexibility than was previously possible with HTML. In older versions of HTML, Web designers could set the background only for the entire page. With the exception of tables, individual HTML elements had the same background color as the page background. You can still apply a color to the entire page, using `body` as the selector, like this:

```
body{margin: 20px 50px 20px 50px;background-color:
    #9CF;}
```

With CSS you can use different background colors for different elements on a page, while using another background color for the page as a whole. Not only can you apply background colors to block-level elements like headings and paragraphs, but you can also apply them to in-line elements (e.g., the `em` tag). If you wanted to highlight your headings, you could assign a particular background color to them, and the code would look like this:

```
h1{font:bold x-large/2 verdana,helvetica,sans-serif;
    color:#f03; background-color:#ff3}
```

A page that used this style rule for `h1` and `#9cf` for a background color for the body would look like Figure 7–1 in a browser (I've also used a navy blue for the paragraphs).

Legible Color Combinations

The right combination of foreground and background colors can make your Web site more attractive and easier to read, while attracting more attention from Web surfers. The wrong combination can make users move on quickly to another site.

The most important factor in choosing colors for text and background is contrast: sharp contrasts make text easier to read, and most research has found that dark characters on a light background are more legible than light characters on a

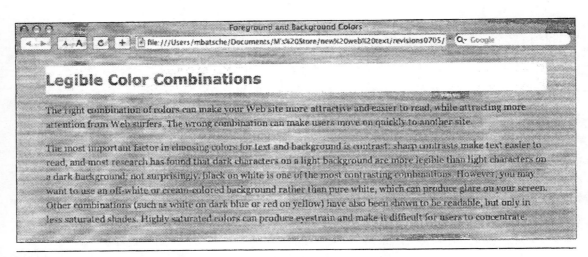

FIGURE 7–1 A page showing background colors

dark background. Not surprisingly, black on white is one of the most contrasting combinations. However, you may want to use an off-white or cream-colored background rather than pure white, which can produce glare on your screen. Other combinations (such as white on dark blue or red on yellow) have also been shown to be readable, but only in less saturated shades. Highly saturated colors can produce eyestrain and make it difficult for users to concentrate.

According to Michael Bernard (2003), research has shown that the *least* readable color combinations include green on yellow, red on green, and fuchsia on blue. As a general rule, it's a good idea to avoid complementary colors—that is, colors located across from each other on the color wheel (e.g., red and green, blue and orange, yellow and violet)—at high saturation. Used in combination, these colors can seem to vibrate and make reading very difficult.

Color-blind users have particular problems with text and background combinations that Web designers need to be aware of. Chuck Newman (2000) provides the following guidelines:

❑ Keep colors bright. Even though color-blind users may not be able to see the colors themselves, bright colors are easiest to tell apart.

❑ Less-safe colors (e.g., red and green) can be made more safe against the right background. Red can be used against white; green or turquoise can be used against black.

❑ Although red and green present problems, red is more easily confused against dark green. Different shades, such as magenta and teal, can work together.

❑ A color scheme that uses various shades of a single color can also create problems for color-blind users since they may have trouble differentiating among the shades.

❑ If color carries any important information (e.g., a link color), it should be supplemented by some other type of marker (e.g., a hover underline).

BACKGROUND IMAGES

You can also place images in the background of your page, like wallpaper on your computer's desktop or watermark pictures in the background of a text page. The same rules hold true for background images as for other images: they must be either GIFs, JPEGs, or PNGs, and they should be small files in terms of both their memory requirements and their dimensions. In previous versions of HTML, these images were "tiled," that is, they were repeated horizontally and vertically, depending on the size and shape of the image. However, CSS again allows you more flexibility in the use of background images. You can still tile them if you wish, but you can also repeat them only once, or you can repeat them vertically or horizontally, depending on your page layout. You can also position your background images more precisely using CSS. In fact, contemporary Web designers are using background images in particularly imaginative ways, in many cases replacing the foreground images used in earlier Web designs (see appendix 2 about image replacement).

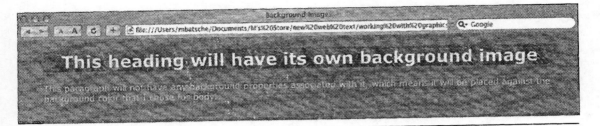

FIGURE 7–2 A heading with a background image

In addition, you can use a tiling graphic behind a particular part of a page; for example, you could have all of your level-one headings placed on a textured background to highlight them. The property is `background-image`, and the code would look like this:

```
h1{background-image:url(background.gif)}
```

Instead of `background.gif`, you would insert the name of the graphic you're using for your background tile. A heading with a background image would look like Figure 7–2.

Background Image Repetition

Besides letting you insert background images, CSS will also let you determine how you want a background image to repeat: horizontally, vertically, or not at all. The property here is `background-repeat`, and you have several choices for values:

❑ `repeat`. If you use this value, the background image will be repeated throughout the background area of the selector, both horizontally and vertically. If you don't specify a `background-repeat` value, `repeat` is the default.

❑ `repeat-x`. If you use this value, the image will be repeated horizontally but not vertically. The graphic will be repeated in one straight, horizontal line behind the selector.

❑ `repeat-y`. If you use this value, the image will be repeated vertically but not horizontally. The graphic will be repeated in a straight vertical line behind the selector. The default position is on the left edge of the containing block.

❑ `no-repeat`. If you use this value, the image will not be repeated; it will appear only once behind the selector.

The code looks like this:

```
body{background-image:url(background.gif); background-
     repeat:no-repeat;}
```

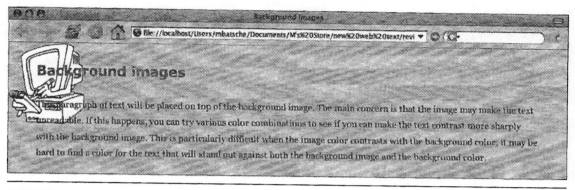

FIGURE 7–3 A nonrepeated background image

Notice that you must begin by specifying the background image you want to use; then you can determine how that image will show up in the browser. In a browser this code would look like Figure 7–3.

Background Image Positioning

If you look at the example in Figure 7–3, you'll notice that the image is placed on the left side of the page at the top. That's the default position for a background image in CSS. However, there may be times when you want the background image somewhere else on the page. Here, for example, the background image makes the composition of the page seem unbalanced. The property to move the background image is `background-position`, and there are several values you can use:

- ❑ `top`. This aligns the image with the top edge of the selector's background. Frequently, you'll use `top` with another dimension, such as `left`, `right`, or `center`.
- ❑ `bottom`. This aligns the image with the bottom edge of the selector's background. Frequently, you'll use `bottom` with another dimension, such as `left`, `right`, or `center`.
- ❑ `left`. This aligns the image with the left edge of the selector's background. Frequently, you'll use `left` with another dimension, such as `top`, `bottom`, or `center`.
- ❑ `right`. This aligns the image with the right edge of the selector's background. Frequently, you'll use `right` with another dimension, such as `top`, `bottom`, or `center`.
- ❑ `center`. This aligns the center of the image with the center of at least one of the axes of the selector's background. `center` can be used with any of the other dimensions; but if it is used alone, it is equivalent to `center center`.
- ❑ *Length value.* If you use a length value, such as 100px or 20%, the image will be offset that much from the top left corner, which is the default background

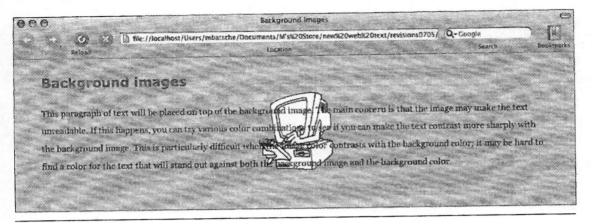

FIGURE 7–4 A positioned background image

image position. If you give only one length value, it will be used for both horizontal and vertical positioning. If you give two length values, the first will be the horizontal position, and the second will be the vertical position.

If I decided to center the background image in Figure 7–3, the code could look like this:

```
body{background-color: #9CF; background-image: url
     (background.gif); background-repeat:no-repeat;
     background-position:center center;}
```

In a browser it would look like Figure 7–4.

Background Image Attachment

There's one more property that you can use with background images, but it's difficult to demonstrate with a screen shot. Background attachment allows you to fix a background image so that the content of the page scrolls over it. Or you can allow the background image to scroll along with the content. There are two values:

- ❏ fixed. This fixes the background image in place so that the content scrolls across it.
- ❏ scroll. This allows the background image to scroll with the content; scroll is the default.

If I wanted the background image in Figure 7–4 to be stationary, I'd add this code:

```
body{background-color: #9CF; background-image:
     url(background.gif); background-repeat:no-repeat;
     background-position:center center; background-
     attachment:fixed;}
```

Background Image Designs

When you choose a background image, your main consideration is making sure the combination of image and text will still be readable. There are several points to keep in mind:

- ❑ If you use a light background, keep your text dark, and vice versa.
- ❑ With patterned background images, try to use patterns that are either all dark or all light colors. Otherwise, your type may show up well on one part of the pattern and poorly on another part.
- ❑ Avoid busy background patterns that interfere with the type. And be sure to test your background image and type to make sure that the combination works.

Another aspect to consider when you choose a background image is file size. A large graphic may look impressive, but if it's too large, it will take too long to download, and your visitors may never see it (the background image loads after the text). It's better to use a small graphic or a graphic that has been compressed to take the file size down.

THE BACKGROUND PROPERTY

CSS provides a way to set all of your background values at once, using the `background` property. You can set all five properties with `background: color, image, repeat, position,` and `attachment`. Unlike the `font` property, these properties do not have to be given in any particular order. However, they must include a color, or the property won't be read. A background rule looks like this:

```
body{background:#9cf url(background.gif) no-repeat
    center center fixed;}
```

EXERCISE: CHOOSE FOREGROUND AND BACKGROUND COLORS

Choose foreground and background colors for your pages. Remember to make the combinations legible. It's also a good idea to maintain consistent colors throughout your site, that is, all the pages at the same level of importance should use the same colors in the same way. You can also try a background image if you find something appropriate.

BORDERS

CSS has added one more graphic effect to HTML: borders. Previously, borders were available only for tables and images; now they can be applied to any element, from an entire page to a single word. Borders can be used to highlight an element or to separate sections of a page layout. However, they should be

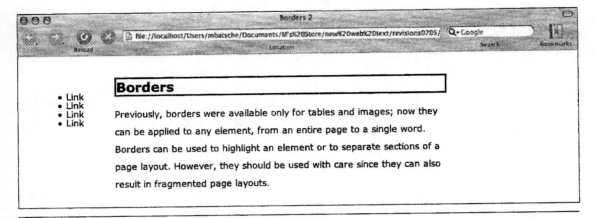

FIGURE 7–5   A full border

used with care since they can also result in fragmented page layouts. The property for a full border is border, and a rule looks like this:

```
h1{border: solid medium #03f}
```

In a browser it would look like Figure 7–5.

Like background, border is a shorthand property that allows you to set all the values of the border at once. However, you can also use border-left, border-right, border-top, or border-bottom if you want to have a border on only one side of an element. For example, if you wanted to have a vertical line separating your menu from your content, you could put a border down one side of the content box. In that case you could use a border rule like this:

```
.content{border-left: solid thin #900}
```

In a browser the result would look like Figure 7–6.

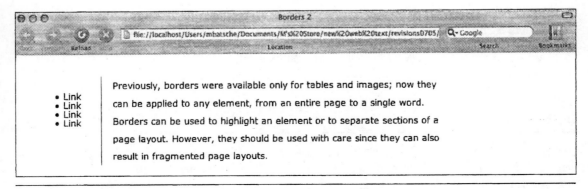

FIGURE 7–6   A border on the left

These are the other border properties:

❑ `border-style`. This sets the appearance of the border. **You must always include `border-style` in your border properties; if it isn't included, the border won't show up (the default is none).** These are the values: none (no border, the default), `dotted`, `dashed`, `solid`, `double` (double line), `groove` (3D groove using colors lighter and darker than those specified in the border color), `ridge` (3D ridge), `inset`, and `outset`. You can set different styles for different sides: (a) one value will apply to all four sides; (b) two values will apply the first value to the top and bottom and the second value to the left and right; (c) three values will apply the first value to the top, the second value to the left and right, and the third value to the bottom; (d) four values will set the top, right, bottom, and left borders. You can also apply the style to a single side by using `border-top-style`, `border-right-style`, `border-bottom-style`, or `border-left-style`. Some of these border styles (i.e., `groove`, `ridge`, `inset`, and `outset`) will be more visible with a wide border width, and `inset` and `outset` do not always show up well in a browser. The examples in Figure 7–7

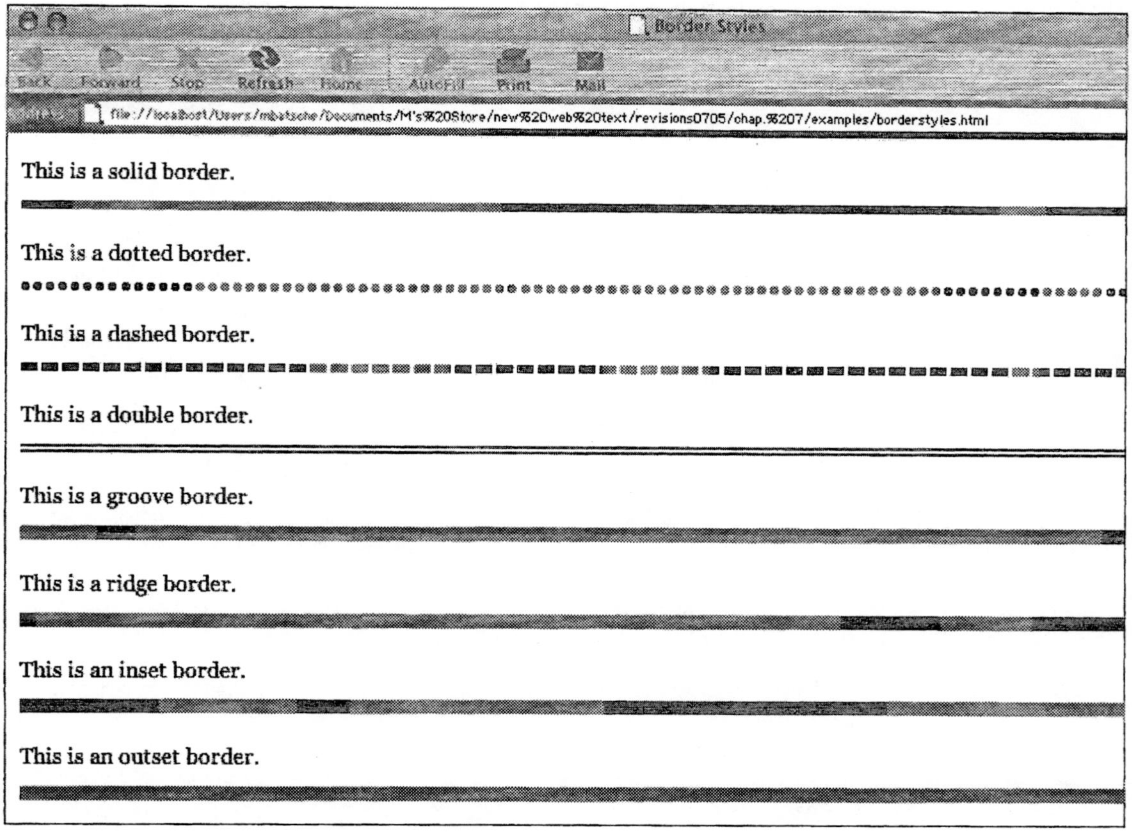

FIGURE 7–7 Border styles

show a border width set to `thick` for `solid`, `dotted`, `dashed`, and `double` and to 10px for `groove`, `ridge`, `inset`, and `outset`.

❏ `border-color`. This sets the color of the border; as usual, you can use any of the color notation systems for the value. Depending on the number of values you use, you can set different colors for different sides. You can also set colors for the individual sides of the border by using `border-left-color`, `border-right-color`, `border-bottom-color`, or `border-top-color`.

❏ `border-width`. This sets the width of the border. The values are `thin`, `medium`, `thick`, or a length value. As with `border-style` and `border-color`, you can set the different sides of the border to different widths. You can also use `border-left-width`, `border-right-width`, `border-top-width`, or `border-bottom-width` to set just one side.

To set all of the border properties at once, the code looks like this:

```
h1{border: solid #c60 thick;}
```

Unlike the other border properties, `border` cannot set individual sides to different styles. To do that, you'll have to use the specific properties (i.e., `border-style`, `border-color`, and `border-width`). Because the default border style is `none`, you must always include a border style in your border values.

PADDING

Padding is the other type of spacing that we mentioned in chapter 5, along with margins. Padding is particularly important (and useful) with borders. Margins go around the outside of the element, whereas padding goes inside the edge of the element box. Since the edge of this box may not be evident unless the box has a border, the difference between margins and padding may not be obvious in most cases. However, when the border has been added, the difference between the two types of space is clear.

Padding uses the same properties and values as margins, that is, `padding`, `padding-left`, `padding-right`, `padding-top`, and `padding-bottom`. The values that can be used with padding are also the same as those with margins:

❏ *A length measurement*—for example, pixels (`px`), centimeters (`cm`), ems (`em`), or inches (`in`)

❏ *A percentage* (padding width as a percent of the total page width)

❏ `auto`, which will set the padding to the default in the browser.

You can apply all of these properties to HTML tags, but you can also use them with CSS classes and ids.

You'll probably want to include padding for any element you place within a border. As you can see from Figure 7–5, it can be difficult to read text that extends all the way to the border edge, and it looks crowded on the screen.

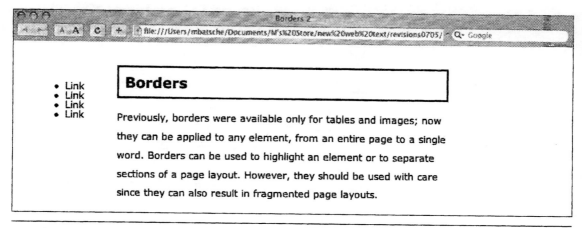

FIGURE 7–8   A border with padding

You can experiment with amounts, but 10 to 20 pixels is usually adequate. Figure 7–8 shows what Figure 7–5 would look like if the style rule were written like this:

```
h1{border: solid medium #03f; padding:10px;}
```

EXERCISE: SET A BORDER

If you have page elements that need to be set off from other elements on the page (e.g., sidebars, menus), try applying a border to the element. Set the border style, color, and width. Try combining a border with a background color for emphasis. Remember to include padding in order to have some space between the content and the border.

PROCESS FOR SETTING COLORS, BACKGROUNDS, AND BORDERS

Use the following steps to style the foreground and background of your pages:

1. Choose colors with high contrast for the text and the background of your Web pages. Use hexadecimal color numbers for most situations.
2. If you want to use a background image, choose one that won't interfere with any text printed across it. Set the image repetition, position, and attachment.
3. Apply borders to HTML elements for emphasis, but don't apply borders to every element, or the page will appear fragmented.

4. If you choose to use a border, remember to include a `border-style` property, or the border will not be visible. Remember to place padding between the element and the border.

CHAPTER SUMMARY

Colors

❏ Colors on the Web can be expressed with one of sixteen predefined colors, with hexadecimal numbers, or with RGB color values. The numeric values are more exact and provide more color possibilities.

❏ Legibility is the most important factor in the choice of color combinations. For long text sections, dark type on a light background is usually considered most legible. Designers must also keep contrast sharp to aid color-blind users.

Backgrounds

❏ Background colors can be provided for any HTML element, using the CSS `background-color` property.

❏ Background images can be added to any HTML element, using the CSS `background-image` property. They can be positioned exactly, their repetition can be specified, and they can be attached or scrolling.

Borders

❏ Borders can be provided for any HTML element, using the CSS `border` property. Designers can choose style, color, and width for the border, but style must be included as a property, or the border will not be visible.

CSS PROPERTIES IN THIS CHAPTER

PROPERTY	VALUE	EFFECT
color	Predefined color names, hexadecimal color value, RGB color numbers	Sets color for an element
background-color	Predefined color names, hexadecimal color value, RGB color numbers	Sets background color for an element
background-image	url(filename.gif)	Provides browser with file name of background image
background-repeat	repeat, repeat-x, repeat-y, no-repeat	Sets the type of repetition for the background image
background-position	top, bottom, left, right, center, length value	Provides location for background image (used in pairs or alone)

Continued

Continued

PROPERTY	VALUE	EFFECT
`background-attachment`	`fixed, scroll`	Sets background image so that text scrolls over it or it scrolls with the text
`border-color`	Predefined color names, hexadecimal color value, RGB color numbers	Sets color of entire border; for a single side use `border-top-color, border-right-color, border-bottom-color,` or `border-left-color`
`border-style`	`none, dotted, dashed, solid, double, groove, ridge, inset, outset`	Sets style of entire border; for a single side use `border-top-style, border-right-style, border-bottom-style,` or `border-left-style`
`border-width`	`thin, medium, thick,` or a length value	Sets width of entire border; for a single side use `border-top-width, border-right-width, border-bottom-width,` or `border-left-width`
`border-top`	Style, color, width	Sets top border only; must include style
`border-right`	Style, color, width	Sets right border only; must include style
`border-bottom`	Style, color, width	Sets bottom border only; must include style
`border-left`	Style, color, width	Sets left border only; must include style
`border`	Style, color, width	Sets entire border; must include style

SOURCES

Bernard, M. 2003. How should text be presented within a Web site? *Criteria for optimal Web design (Designing for usability).* http://psychology.wichita.edu/optimalweb/text.htm.

Clark, J. 2002. *Building accessible Websites.* Indianapolis, IN: New Riders.

Newman, C. 2000. Considering the color-blind. *New Architect* (August). http://www.webtechniques.com/archives/2000/08/Newman/.

Linking Pages

LINKS

According to usability expert Steve Krug (2000), "Navigation isn't just a feature of a Web site; it is the Web site. . . . Without it, there's no there there" (59). It's undeniable that a good navigation system can mean the difference between satisfied users, who are likely to return to your site, and unhappy users, who leave your site feeling frustrated and confused. The question is, How do you make your navigation reflect the structure of your content while also making it easy for your user to get around your site? In this chapter we'll discuss some basic principles of navigation and linking. We'll discuss the major types of links: relative, absolute, and on-page anchors. In addition, we'll go over the CSS properties that you can use to apply colors and styles to various link states.

NAVIGATION

It's easy to see the resemblance between navigating through a Web site and searching through a real physical location. Normally, you arrive at a Web site looking for something, and you either browse through the options listed, or you use a search tool if one is provided. Jakob Nielsen (2000) divides Web users into "search-dominant" and "link-dominant" groups (224). Search-dominant users immediately go for the search options; link-dominant users browse the site, returning to search if they can't find what they're looking for in any other way. "Mixed-behavior" users switch back and forth between a search and links, depending on which approach seems most promising on a given site. The decision to browse or search depends on factors like the urgency of finding the information and the amount of time available to search, but it can also depend on whether the site's navigation seems usable.

The purpose of navigation goes beyond simply providing a means of getting to a particular point. As Krug (2000) points out, "Done right, navigation puts ground under our feet (even if it's virtual ground) and gives us handrails to hold on to—it makes us feel grounded" (59). Navigation makes the major sections of a site's content visible, revealing the structure of the site and indicating how to move around and how to begin searching.

Global Navigation

Every page in your site should contain the site's global navigation—that is, links to the site's major sections. Moreover, that navigation should appear in the same place on every page or at least in the same place on every page that's at the same level of importance in the site's hierarchy. Krug (2000) defines the minimum content for global navigation, which should appear everywhere (62):

- ❏ A site ID (i.e., a logo or a site identifier of some kind)
- ❏ A way home (i.e., a home button)
- ❏ A way to search
- ❏ Utilities (e.g., help sections, frequently asked questions, shopping baskets)
- ❏ Site sections (the major parts of the site)

Figure 8–1 shows the home page from the "Missions" section of the NASA site, which includes Krug's minimal navigation content. The NASA site is large scale, but each main section is indicated in the horizontal menu bar below the page header. Then the home page of each main section indicates the subsections of that section.

FIGURE 8–1   NASA missions page

Having a site ID on every page guarantees that users never have to wonder if they've wandered off your site by mistake—the ID always reminds them of where they are. The site logo is most frequently placed at the top of the page, and increasingly it's linked to the site's home page. As Patrick Lynch and Sarah Horton (2001) point out, "Unlike designers of print documents, designers of Web systems can never be sure what other pages the reader has seen before linking to the current page" (94). The site ID ensures that users are never in doubt about what site they're currently visiting.

You might think that having a logo linked to your home page would do away with the need for a separate home button, but there will always be a user who isn't familiar with the convention or who forgets that clicking on the logo will take him or her to the home page. A home button is extra reassurance.

Although you probably won't be responsible for developing a search utility for your site because that requires advanced programming skills, having a search available on your site will make it much easier to use. Nielsen's research (2000) has shown that over half of all users fall into the search-dominant category. Therefore, not surprisingly, Nielsen argues, "Search should be easily available from every single page on the site" (225). Some search engines offer free versions for nonprofit groups, or you may choose to pay for a commercially developed search program.

What constitutes a utility varies from site to site. Krug (2000) describes utilities as "links to important elements of the site that aren't really part of the content hierarchy" (65). Utilities can include frequently asked questions, contact information, help screens, and site indexes. On the NASA site the utilities include a text-only version of the site, a version that doesn't use Flash animation, and a contacts page.

The section labels that you include in your global navigation will be the labels you developed when you were working out the organizational structure for your site; they're on the highest level of your hierarchy after your home page. Whether you include any subsection labels, as the NASA page does, goes back to the question of breadth versus depth. A broad, shallow structure will have more choices at higher levels and fewer categories to click through. A more narrow, deeper structure will have fewer choices at the higher levels of organization, but more categories to browse through.

Links and URLs

Technically, a **link** is a connection between two Web resources with anchors at either end. As the W3C specification states, "The link starts at the 'source' anchor and points to the 'destination' anchor, which may be any Web resource (e.g., an image, a video clip, a sound bite, a program, an HTML document, an element within an HTML document)." In fact, the img tag is also a link since it points to a Web resource, that is, an image. You can have anchors between pages, but you can also have anchors within a single page, such as the links that allow you to click back to the top of a long text passage without scrolling. At the base of most links is either an absolute or a relative **URL** (uniform resource locator).

SIDEBAR: LINK VOCABULARY

Absolute link: A link in which all parts of the URL are specified; usually used for links that go off a site.

Anchor: The point at which a link originates and ends. The page with the link is the *source anchor;* the resource to which the link points is the *destination anchor.*

HTTP: Hypertext transfer protocol, the protocol by which Web pages are sent and received.

Path (path name): A sequence of symbols and names that identifies the location of a file. The simplest path name is just a file name, which directs the browser to look for the file in the current folder. However, if the file resides in a different folder, the path tells the browser where to find that folder and then the file. The path begins from the file and then works up to a higher level or down to a lower one, separating the names of folders with forwardslashes.

Relative link: A link in which only part of the URL is specified; used for links within a site.

URL: Uniform resource locator; also called URI, or uniform resource indicator. The address giving the location of resources on the Web, including pages, video files, sound files, and so on.

URLs are addresses for Web pages; they indicate the location of a given page or a given file. In other words, the URL tells the browser where to look to find a particular Web resource. The parts of the address are separated by forward slashes, and each part conveys different information to the browser. Here's a typical URL and an explanation of its parts:

```
http://communication.utsa.edu/mbatch/default.html
```

- ❑ http—http tells the browser what protocol to use when it requests the page from a server (http stands for hypertext transfer protocol). Most Web pages use http rather than other protocols like ftp or gopher.
- ❑ communication.utsa.edu—This section of the URL is called the *domain;* it tells the browser what server to contact for the page. In this case the name of the server is communication, and it's located at utsa, the University of Texas at San Antonio, which is an educational institution— hence the .edu domain extension.
- ❑ /mbatch/default.html—The rest of the address gives the *path* that the browser must take to find the particular page requested on the specified server. mbatch is a directory (i.e., a folder) found on the communication server. The order of directories in the URL goes from higher (i.e., more inclusive) to lower—thus, communication includes the mbatch directory, which includes the file default.html. The final part of the URL, default.html, is the actual HTML file the browser is to read.

This particular URL is *absolute*, that is, it tells the browser every part of the address so that the browser can locate exactly that page. However, there are times when you don't have to use the entire URL—just a part of it. When you're linking together the pages you've created on your own site, you usually don't need to supply the protocol and domain names because the browser is simply moving from one page to another in the same domain, using the same protocol. These shortened URLs are called *relative* URLs, because they're related, or connected, to other pages on the same site. For example, in the site at the URL just listed, which is my site at my university, I have a link to another page, `classes.html`, which is part of the same Web site as `default.html`. If I write a link to this second page from `default.html`, the URL is written as `classes.html` rather than `http://communication.utsa.edu/mbatch/classes.html` because its address is relative to the address of the page the browser has already opened. By linking to `classes.html`, I'm telling the browser to look for this page in the same folder where it found `default.html`. We'll talk more about relative URLs later on in this chapter.

Relative URLs are used to move from one page of a site to another and to locate resources like images; an attribute and a value like `src="logo.gif"` is actually a relative URL. They may include the names of folders along with the name of the file. For example, if I kept all of my images together in a folder called `images`, then the relative URL for one of these images would be `/images/logo.gif`. Relative URLs won't include all the parts of the absolute URL. You'll use an absolute or a relative URL when you write most of your links. The only exception will be when you write links between anchors on the same page.

MAKING LINKS

The HTML for both absolute and relative links is basically the same, although the URLs are written differently. However, as with your image tags, when you create a link, **you must type the URL *exactly* as it's written on the page you're linking to,** including capitalization and spelling. When you're surfing the Web, most browsers will allow you to type a part of the URL into the address space and have the browser fill in the rest. You can't do that when you write a link. If it's an absolute link (i.e., a link to another Web site), you must include the entire URL (starting with `http://`) and use exactly the same spelling and capitalization.

Absolute Links

You saw the basic link tag in chapter 2: `<a href="url"> </a>`. The difference comes in what you type into the URL space. With absolute links you must use the entire URL, including protocol, server, and path. If I wanted to link to the Webopedia site, an online encyclopedia of Web terminology, my tag would read `<a href="http://www.webopedia.com">`. I don't need to include an HTML file name here because the domain name designation, `www.webopedia.com`, links to the entry page of the site.

You must be careful to write the domain name exactly as it's given in the site URL. If you've ever spent time trying to guess what the correct URL is for a site without knowing the exact domain name, maybe typing in several variations, you know how important it can be to get the domain name exactly as it appears on the page to which you're linking. In fact, it's frequently easiest to simply go to the site you want to link to, copy the URL in the address field of your browser, and paste it into your link.

Clickable Text

Between the `<a href="url">` and the `</a>`, you'll write whatever you want your users to click on to activate your link. You can use either text or an image for your link, although there are some drawbacks with images that we'll discuss later in this chapter.

In most browsers the default formatting for links is underlining, with blue text for unvisited links and purple text for visited links. You can change those specifications using CSS, but it's a good idea not to use underlining for anything other than links on your site. By now the convention of underlined links is so widespread that users are likely to assume that any underlined text is linked text.

Linking to E-mail

Web sites should include an e-mail address for someone associated with the site who can serve as a contact person: the president of the organization, the manager of the company, the sales office for the product. If nothing else, most pages include the e-mail address of the designer, frequently in a separate address section at the bottom of the first page.

Linking to an e-mail address uses a format very similar to absolute links but with one major difference: instead of using an absolute URL, you insert

---

SIDEBAR: CLICKABLE TEXT TIPS

Here are some things to keep in mind when you're creating your clickable text:

- ❑ Keep the text to be clicked on relatively short—it's hard to click on large segments of text.
- ❑ Don't use "click here" for your text. Users need to have some idea of where they're going to go if they click on the link. Moreover, "click here" will be meaningless for screen readers used by those with vision disabilities and in Web-enabled devices.
- ❑ If you use italics or boldface on the clickable text, be consistent from one link to another. Don't italicize some links but not others, or you'll confuse your readers.

"mailto:designer@emailaddress.com", with your e-mail address placed after the colon. Your clickable text follows the e-mail address. The code looks like this:

```
<a href="mailto:designer@emailaddress.com"> Designer's
    e-mail address</a>
```

A page with both absolute and e-mail links looks like this:

```
<!DOCTYPE HTML PUBLIC "-//W3C//DTD HTML 4.01
    Transitional//EN" "http://www.w3.org/TR/1999/
    REC-html401-19991224/loose.dtd">
<html>
<head>
    <title>Absolute Links</title>
    <link rel="stylesheet" href="styles.css"
    type="text/css">
</head>
<body>
<h1>A Page with Links</h1>
<p>
Most pages have both absolute and relative links.
The absolute links link one Web site to another.
The relative links link individual pages within
a site.
</p>
<p>
Absolute links must use the entire URL of the page
that you're linking to. Some people are used to typing
in only a part of the URL—for example, google—and
then letting the browser fill in the rest—for example,
http://www.google.com/. But when you write HTML,
whether it's in an authoring program or an HTML
editor, you must include the entire URL, including
protocol and domain. The following is an absolute
link, including the entire URL:< br>
<a href="http://communication.utsa.edu/mbatch/
    default.html"> Batschelet Web site</a>
</p>
<p>
E-mail links are also absolutes, but the format is
slightly different. Here's an e-mail link:<br>
<a href="mailto:margaret.batschelet@utsa.edu">
    Batschelet e-mail</a>
</p>
</body>
</html>
```

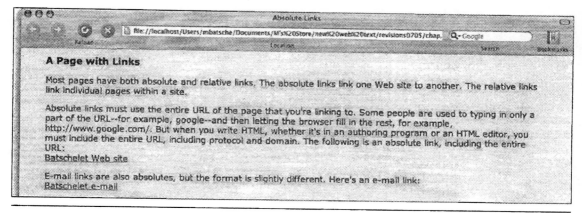

FIGURE 8–2   Absolute links

And in a browser it would look like Figure 8–2.

Relative Links

The other common type of link is the relative link. The HTML uses the same container tags, so a relative tag would look like this: `<a href="index.html">Back to Home page</a>`. That code tells the browser to look for the file `index.html` in the same folder that is currently showing. Occasionally, you'll need to add a subfolder name to the URL path if you have subfolders in a larger folder. For example, if I wanted to link to a page called List in a subfolder called `stylebooks`, the link tag would read `<a href="stylebooks/list.html">List of Stylebooks</a>`. This code tells the browser to look for a subfolder called `stylebooks` in the same folder containing the current page and to find the page called `list.html`. If I wanted to move up one level to the next level of folders in a site, I'd use `../`. For example, if I wanted to link the `list` page to a page in the larger folder, I'd write the code like this: `<a href="../index.html">Back to Home</a>`.

The code for relative links looks like this:

```
<!DOCTYPE html PUBLIC "-//W3C//DTD XHTML 1.0
    Transitional//EN" "http://www.w3.org/TR/xhtml1/
    DTD/xhtml1-transitional.dtd">
<html xmlns="http://www.w3.org/1999/xhtml">
<head>
    <meta http-equiv="content-type"
    content="text/html; charset=utf-8" />
    <title>Relative Links</title>
    <link rel="stylesheet" href="styles.css"
    type="text/css">
```

```
</head>
<body>
<h1>A Page with Relative Links</h1>
<p>
Most pages have both absolute and relative links. The
relative links link a page within one Web site to
another page within the same site.
</p>
<p>
Relative link URLs have only the name of the file and
any path information needed for the browser to find
the page. For example, if I wanted to link to another
page in the same folder that this page is in, the
relative link would look like this:<br />
<a href="page2.html">Another page in the
    same folder</a>
</p>
<p>
If I wanted to link to another page in a subfolder
called "Spring" (located in the same general folder),
the relative link would look like this:<br />
<a href="spring/communicationclasses.html">Another
    page in a subfolder</a>
</p>
<p>
On the other hand, if this page were located in a
subfolder and I wanted to link to a page in the main
folder, the link would look like this:<br />
<a href="../allclasses.html">Another page in the main
    folder</a>
</p>
</body>
</html>
```

And Figure 8–3 shows what it would look like in a browser.

You may notice that the links in Figure 8–3 don't look much different from those in Figure 8–2. In a browser, links all look the same, whether they're absolute or relative. The only real difference comes at the code level.

Using Images for Links

You can also use pictures as your clickable objects. The HTML for image links is the same as that used for text links; the only difference is that you put an image tag where you previously put clickable text. The result looks like this:

```
<a href="page2.html"><img src="Sandstone-home.gif"
    alt="return to home" width="64" height="24" /></a>
```

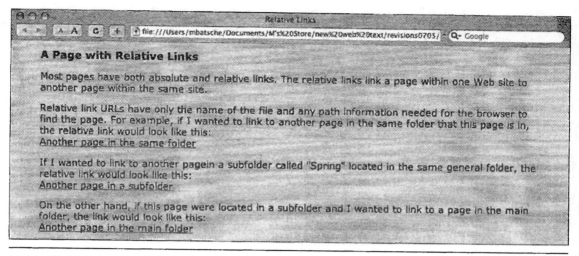

FIGURE 8–3   Relative links

EXERCISE: CREATE LINKS

By now you should have more than one page created for your site. Write some relative links to link the pages together. Add a link to your e-mail address and any absolute links to other sites that you're going to use.

In a browser it would look like Figure 8–4.

Notice that the image link has a border around it in the same color as the link color. (To remove the border in some browsers, use `img` as a selector in your style sheet, and specify `border-style:none`. The rule looks like this: `img{border-style:none;}`). The only indication that an image link has

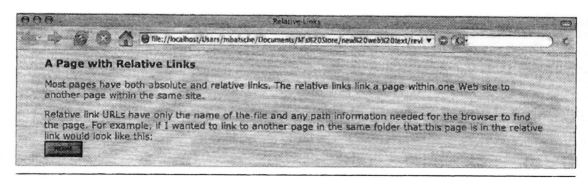

FIGURE 8–4   An image link

been visited is that the border changes color on visited links. That's one of the problems with image links: it's more difficult to indicate whether a link has been visited, something most users want to know. Another problem is the amount of work it takes to keep image links up-to-date. If you had a page called "Activities" and you decided to change the name to "Current Events," you could simply change the clickable text on the pages where the links occurred if your links used text. But if your links used images, you'd need to create a new image, usually a more time-consuming activity than just retyping the page name. And on a large site with many links at many levels, these maintenance problems can be acute.

Here are some other points to keep in mind with image links.

- ❏ Keep the linked images clear and unmistakable. Although your users can identify linked images by moving their cursors across them, they may not do so if they don't understand in the beginning that the images are linked.
- ❏ Consider using words along with your image links. Links using words clearly change color after they've been visited and are less likely to be mistaken for something else.
- ❏ Consider using a combination of words for some links and images for others, perhaps putting your main navigation in words but using images for special purposes like help or search, which aren't likely to change.
- ❏ Be particularly careful to include `alt` attributes identifying the destination of each link; navigation is one thing you don't want users with screen readers to miss!
- ❏ Remember that the Web is an international medium; icons that seem perfectly clear to a reader from the United States may be confusing to a reader from another country. This is another reason to keep your main navigation in words.

EXERCISE: LINK IMAGES

If you want to use linked images, try inserting them now. Remember to make them clear and unmistakable as links.

On-Page Links

One final type of link can be used when you need to link information on a single page. If you have a long page with a variety of information on it, your users may want to skip some items and go directly to others. You can provide them with a link list at the top of the page that allows them to go directly to topics of interest.

Absolute and relative links both refer to the URLs of other pages, whether off-site or on-site. Since on-page links link to material on the same page, you don't need a URL. Instead, you create an on-page anchor, a line of code on the page, and then refer to that anchor in your link. There are two ways to set up on-page anchors: using the `name` attribute or the `id` attribute.

Anchors Using the `name` ***Attribute.*** The HTML attribute and value for an on-page anchor are `name="anchorname"` and are used with the `<a>` tag.

The anchor name can be anything (although it can't begin with a number), but for the sake of simplicity, it's usually easiest to simply insert the anchors next to any headings you use and then use the wording of the headings for the anchor names. You insert the anchor name within an anchor tag like this: `<a name="anchor name">`. Because it's inside the angle brackets, the anchor name is invisible and can simply repeat the heading. If I had a heading like "Foundations of Communication," I'd make the anchor `<a name= "foundations">`. I'd put the close tag for the anchor, `</a>`, on the other side of the heading and nest the anchor inside any other tags, like this:

```
<h2><a name="foundations">Foundations of
    Communication</a></h2>
```

This on-page anchor will be invisible in most browsers. However, if the link formatting shows up in the browser (e.g., if the heading is underlined), you can use `border-style:none`, or you can use the ID method instead.

Anchors Using the `id` Attribute. An even easier way of creating an on-page anchor is to use the `id` attribute. You don't have to use the `<a>` tag in this case; you can simply insert the `id` into the existing tag you want to link to. Our previous example would become the following:

```
<h2 id="foundations">Foundations of Communication</h2>
```

The usual `id` rules apply here: you can use a particular `id` only once on a page. But since you'd want each point you're linking to on the page to have a distinct name, that shouldn't present any problem.

Anchor Links. You can link to anchors on the same page, and you can link to an anchor on another page of your site if you want to go directly to a particular area on that page. (The default place for a relative or absolute link to open is the top of the page.) You can even link to anchors on someone else's Web site if you know what the anchor names are (you can find them by looking at

SIDEBAR: ANCHOR NAME AND ID TIPS

There are a few things to keep in mind as you create your anchors:

- ❑ Insert anchors as you write the HTML for your page, attaching them to section titles or other prominent points on the page.
- ❑ Place the anchor next to the text you want to link to; in that way the text will appear on the screen as soon as the user clicks the link.
- ❑ Break up a long Web page by putting anchors on each section and referring to them in a table of contents at the top of the page.
- ❑ Put a return-to-top-of-page link at the bottom of the page or at the bottom of each section.
- ❑ Remember that anchor names can't begin with numbers or underscores.

View Source Code or View Code in your browser). Anchor links use the same basic container tags that other types of links use, with the pound sign referring to the anchor, like this:

```
<a href="#meetings">Meeting Times</a>
```

This code will link to an anchor called "meetings" somewhere on the same page.

When you link to an anchor on another page, you place the anchor name after the file name of the page. In the following example, I'm linking to a page called "Activities" and to an anchor called "meetings":

```
<a href="activities.html#meetings>Meeting Times</a>
```

Here's a longer page with several anchors. Notice that I've also included "Top of page" anchors so that readers can go back to the menu of anchors if they want to make another selection. However, I haven't used *top* as the id because the word *top* is used in some scripting applications and might cause errors. The `<hr>` tag calls for a horizontal rule.

```
<!DOCTYPE html PUBLIC "-//W3C//DTD XHTML 1.0
    Transitional//EN" "http://www.w3.org/TR/xhtml1/DTD/
    xhtml1-transitional.dtd">
<html xmlns="http://www.w3.org/1999/xhtml">
<head>
    <title>Spring 2005 Communication
    Courses</title>
    <link rel="stylesheet" href="styles.css"
    type="text/css" />
</head>
<body>
<h1 id="highest">Spring 2005 Communication
    Courses</h1>
<p>
This is a partial listing of courses available in
Communication for Spring 2005. For more information
consult the bulletin board outside the main
Communication office.
</p>
<ul>
<li><a href="#foundations">COM 2213: Foundations of
    Communication</a></li>
<li><a href="#mass">COM 2343: Introduction to Mass
    Communication</a></li>
<li><a href="#editing">COM 2433: Editing</a> </li>
</ul>
<hr>
<h2 id="foundations">COM 2213: Foundations of
    Communication T/R 12:30-1:45<br />
    Instructor: Dr. Steve Levitt</h2>
```

```
<h2>Readings</h2>
<p>
Course readings available as reading packet; occasional
library reserve materials.
</p>
<h2>Course Description</h2>
<p>
This course will introduce majors to the discipline of
communication and to the communication program at
UTSA. Students will be introduced to fundamental
concepts in speech communication, technical communica-
tion, electronic media, and public relations. Students
will develop essential skills, including the basic
strategies and technologies used for information
access, retrieval, and processing.</p>
<h2>Course Objectives</h2>
    <ul>
    <li>To ensure students have a good understanding
of the principles of human communication as they
apply to the major concentrations of the UTSA degree
program </li>
    <li>To develop library and Internet research
    skills </li>
    <li>To develop writing skills, especially as they
pertain to different academic styles </li>
    <li>To develop skills in the use of information
technologies (e-mail, presentation tools, the
Internet, and World Wide Web)</li>
</ul>
<p>
```
`<a href="#highest">Top of page</a>`
```
</p>
<hr>
```
`<h2 id="mass">,<br />`
```
COM 2343: Introduction to Mass Communication
T/R 2:00-3:15 p.m. Sec. 001 1604 Campus;
Sec. 901 Downtown Campus<br>
Instructor: Dr. Kent Wilkinson
```
`</h2>`
```
<h2>Text</h2>
<p>Branston, Gill, and Stafford, Roy. (2003). <cite>The
media student's book</cite> (3rd ed.). New York:
Routledge.</p>
<h2> Course Description</h2>
<p>
This course encourages students to thoughtfully
examine the influence of mass communication in
```

```
contemporary society. Course requirements include two
midterm exams, a final exam, and online discussion
assignments. </p>
<h2>Course Objectives</h2>
<ul>
     <li>Raise students' awareness of their own media
use and the importance of media literacy</li>
     <li>Familiarize students with key theories that
guide many of our assumptions regarding mass media
effects</li>
     <li>Sharpen students' critical thinking skills</li>
     <li>Clarify the influences of mass communication
in students' fields of study and anticipated future
professions</li>
</ul>
<p><a href="#highest">Top of page</a></p>
<hr>
<h2 id="editing">,<br /> COM 2433: Editing, MWF,
8-8:50<br>
Instructor: Dr. Margaret Batschelet
</h2>
<h2>Texts</h2>
<p>Rew, Lois. <cite>Editing for Writers; Chicago
Manual of Style,</cite> 15th edition.
</p>
<h2>Course Description</h2>
<p>
Communication 2433 (cross-listed as English 2433)
is a course in copyediting, preparing an author's
manuscript for print by making necessary revisions
in mechanics, diction, syntax, format, and
design.</p>
<h2>Requirements</h2>
<p>Students will edit several actual manuscripts
using both traditional editing symbols and MS Word.
In addition, there will be class exercises and
reading assignments. For extra credit, students can
turn in errors that they find in other printed
sources. The class has a midterm and a final. Grades
will be based on the total number of points amassed
during the semester. </p>
<p><a href="#highest">Top of page</a></p>
</body>
</html>
```

In a browser the top of the page (with the on-page menu) would look like Figure 8–5. If I clicked on the third link, the screen would look like Figure 8–6.

Spring 2005 Communication Courses
Spring 2005 Communication Courses
This is a partial listing of courses available in Communication for Spring 2005. For more information consult the bulletin board outside the main Communication office.
• COM 2213: Foundations of Communication
• COM 2343: Introduction to Mass Communication
• COM 2433: Editing

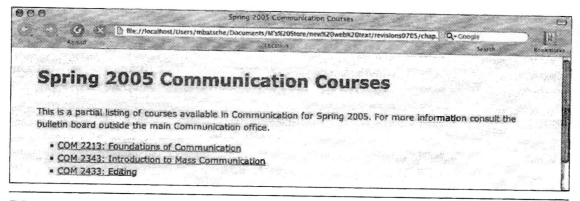

FIGURE 8–5 An on-page menu

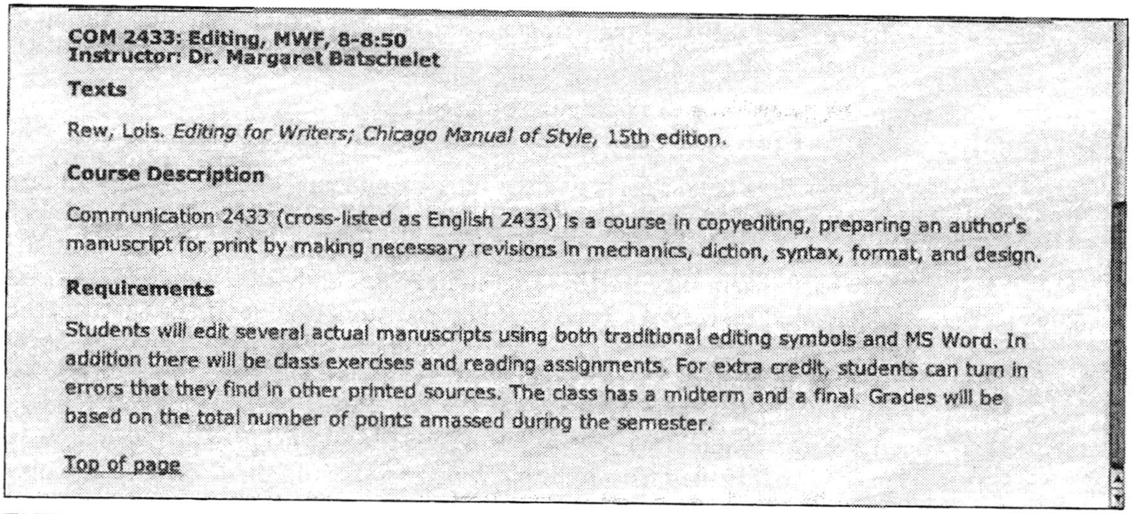

COM 2433: Editing, MWF, 8-8:50
Instructor: Dr. Margaret Batschelet
Texts
Rew, Lois. Editing for Writers; Chicago Manual of Style, 15th edition.
Course Description
Communication 2433 (cross-listed as English 2433) is a course in copyediting, preparing an author's manuscript for print by making necessary revisions in mechanics, diction, syntax, format, and design.
Requirements
Students will edit several actual manuscripts using both traditional editing symbols and MS Word. In addition there will be class exercises and reading assignments. For extra credit, students can turn in errors that they find in other printed sources. The class has a midterm and a final. Grades will be based on the total number of points amassed during the semester.
Top of page

FIGURE 8–6 An on-page link

EXERCISE: ON-PAGE LINKS

Look over the pages you've created. Are any of them longer than one or two browser windows? Consider placing some on-page anchors throughout the page so that your visitors can find information more easily. Place links to the anchors at the top of the page, and place "Return to top" links at the bottom of sections.

LINK ATTRIBUTES

title

The `title` attribute does various things when used with the `<a>` tag, depending on the browser being used. Many browsers display the content of the `title` attribute in a tool-tip box when you move your cursor over the link; some browsers use the `title` attribute for a user's bookmark. The `title` attribute can be particularly useful with image links because it can specify the link's destination (but not whether the link has been visited before). The value for `title` is usually a short text description. The code looks like this:

```
<a href="index.html" title="return to home"><img
    src="house.gif" alt="return to home"></a>
```

If I ran my cursor across the image, the title would appear like Figure 8–7.

target

The `target` attribute comes from the attributes for frames and displays the link destination page in a separate window. The code looks like this:

```
<a href="http://communication.utsa.edu/"
    target="_blank"> Communication Department</a>
```

The idea here is that users will see the new site in a separate window, rather than leaving your site to go to another site when a link is clicked. When they want to return to your site, they simply close the window and are automatically back where they started. This strategy has both pluses and minuses. It's widely used on the Web because it keeps visitors from leaving a site, and the prevailing wisdom is that the longer visitors remain on a site, the more likely they are to return. There's some truth to this, of course, and some visitors appreciate being able to return to the site they started from without having to click back through their history. However, the downside is that visitors may not realize that they have multiple windows open on their screen. They may continue in the new window or even return to your site to visit another link and open another window. In fact, Jakob Nielsen (2000) rates "opening

FIGURE 8–7   A linked image with `title` attribute

multiple browser windows" as one of his Top Ten Mistakes in Web Design. Nielsen argues,

> Designers open new browser windows on the theory that it keeps users on their site. But even disregarding the user-hostile message implied in taking over the user's machine, the strategy is self-defeating since it disables the *Back* button which is the normal way users return to previous sites. Users often don't notice that a new window has opened, especially if they are using a small monitor where the windows are maximized to fill up the screen. So a user who tries to return to the origin will be confused by a grayed out *Back* button. (http://www.useit.com/alertbox/9605.html)

In addition to this problem, the `target` attribute has been deprecated in XHTML, although the same effect can still be achieved using JavaScript. All in all, even though the `target` attribute is still widely used, there are more arguments against using it than reasons for doing so.

USING CSS TO COLOR LINKS

Link color is yet another area in which CSS has provided more options than were originally available in HTML. As I mentioned before, users appreciate having links change color when they've been visited, but that's not the only thing CSS allows you to do. Technically, the link selectors are CSS *pseudoclasses*—they allow you to define styles for a tag *state* rather than just a tag. These pseudoclasses are created like regular classes, but there are two differences: pseudoclasses are attached to the anchor tag name with a colon, and they have predefined names (i.e., you can't choose your own names as you can with regular classes). There are four pseudoclasses associated with links: `a:link`, `a:visited`, `a:hover`, and `a:active`.

Incidentally, if you choose to add all four of these pseudoclasses to your style sheet, you should apply them in the order just given. If you use another order, they may not be applied consistently in all browsers because one style may override another. Dave Shea and Molly Holszchlag (2005) recommend using the mnemonic LoVe/HAte to remember the order.

a:link

The `a:link` pseudoclass is used with unvisited links. You should define a color and a style for your links that will make them stand out in contrast to the text around them. In general, users assume that any contrasting text is link text. The default style in the HTML specifications is that links are underlined and colored blue, which does make the links stand out. However, the default style has some problems. You want your unvisited links to stand out sharply so that it's clear they haven't been visited, but blue is on the cool end of the color spectrum and is thus less likely to catch attention. In addition, even though underlining is an attention-getter, it can also make text difficult to read by cutting off the lower

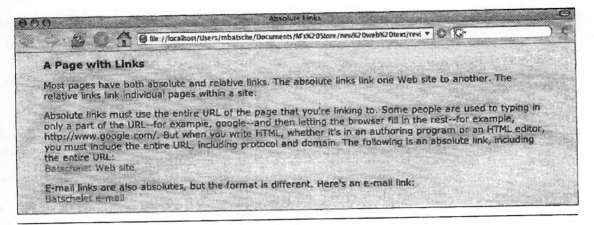

FIGURE 8–8 An `a:link` example

part of letters. So you'll probably want to use the `color` property to change the color of your link and the `text-decoration` property to remove the underlining. The code might look like this:

```
a:link{color:#f39; text-decoration:none;}
```

If I add this style rule to the absolute link page I used in Figure 8–2, it would look like Figure 8–8.

a:visited

Using a different color to indicate that a link has been visited is one of the easiest ways to improve your site's usability. In fact, *not* doing so is another one of Jakob Nielsen's Top Ten Mistakes in Web Design. If users can see where they've been, they can avoid revisiting some pages and return to helpful links. Using a lighter, duller shade of the color you used for your unvisited links helps to show the relationship between the two states, but you need to make sure that the two shades are different enough to be distinguishable. And both shades must show up well against your background color. You don't want a link to disappear just because it's been visited! The important thing is that the visited color be less noticeable than the unvisited color, so that unvisited links stand out more. The code looks like this:

```
a:visited{color:#f96; text-decoration:none;}
```

In a browser it would look like the example in Figure 8–9.

a:hover

The `a:hover` pseudoclass allows you to give feedback to your users when they move their cursors across the screen. A link with an `a:hover` rule changes color or style when the cursor passes across it, indicating that the link

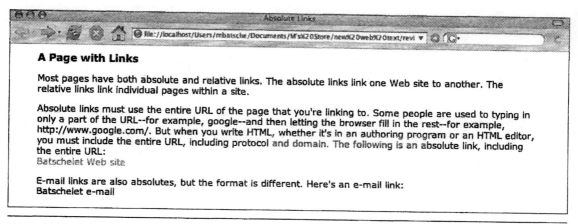

FIGURE 8–9 An a:visited example

is "live." For example, you can use the a:hover selector to use underlining, a style that most Web users associate with links. Even though you may take the underlining off your visited and unvisited links, you can use a:hover to show the underlining when the cursor passes over a link. You can also change colors or even font styles (e.g., adding boldface or italic) if you wish. The code looks like this:

```
a:hover{color:#c0c; text-decoration:underline;}
```

In a browser it would look like the example in Figure 8–10.

Note: CSS2 allows the :hover pseudoclass to be applied to other elements besides anchors, opening up interesting possibilities for drop-down menus and other sophisticated effects. Currently, only the most advanced browser versions

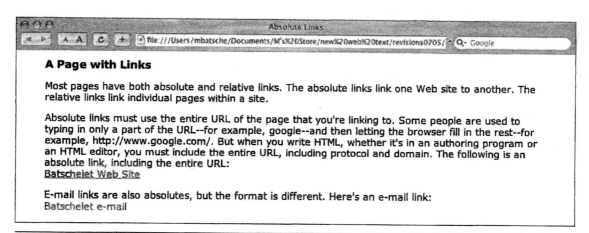

FIGURE 8–10 An a:hover example

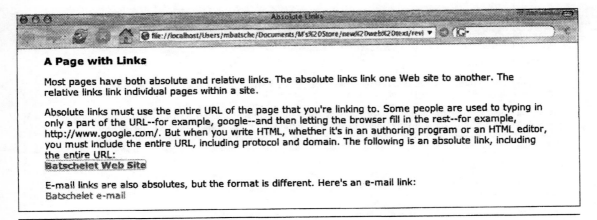

A Page with Links

Most pages have both absolute and relative links. The absolute links link one Web site to another. The relative links link individual pages within a site.

Absolute links must use the entire URL of the page that you're linking to. Some people are used to typing in only a part of the URL--for example, google--and then letting the browser fill in the rest--for example, http://www.google.com/. But when you write HTML, whether it's in an authoring program or an HTML editor, you must include the entire URL, including protocol and domain. The following is an absolute link, including the entire URL:
Batschelet Web Site

E-mail links are also absolutes, but the format is different. Here's an e-mail link:
Batschelet e-mail

FIGURE 8–11 An a:active example

support this effect, but it provides intriguing possibilities for the future. For more information about the uses of :hover, see Shea and Holzschlag (2005).

a:active

The a:active pseudoclass changes the appearance of links as they're being clicked. You can change the link color or style to give users immediate feedback when they click the mouse button. The code looks like this:

```
a:active{color:#930; text-decoration:none}
```

In a browser it would look like the example in Figure 8–11. The box around the link indicates that it's being clicked.

EXERCISE: ADD LINK STYLES

Use your style sheet to style the links that you've added to your pages. Add styles for a:link, a:visited, a:hover, and a:active.

LINK CONCERNS

Apart from the actual mechanics of creating links, there are other questions to be answered: What should you link to, and how should you integrate your links into your page design?

Links as Distractors

It's obvious that links should stand out: if readers can't recognize a link as a link, they won't click on it. But the very fact that links must stand out means that they will contrast starkly with the surrounding text—Web designer Jeffrey Veen (2001) refers to links as "little blue scars." That's not necessarily bad, but a page of text that's peppered with links in a contrasting color may be hard to read and may tempt readers to click away rather than finish reading.

One solution is to group your links at the end of the page or in a column along the side. This doesn't mean that you won't use any links within your text—it's helpful to indicate when other pages provide related information. However, you'll need to find a balance between text that's full of links and text that has no links other than a link to home. You want your readers to be able to move around, following their own interests as well as yours.

When you create an in-text link, make sure that the word you use as the link is a clear indicator of where the link will take the reader. If I had a link like "The *Foundations* class is a prerequisite for Language and Communication Theory," I'd expect a link to a description of the Foundations class. If the link was "The Foundations class is a prerequisite for *Language and Communication Theory*," I'd expect a description of the Language and Communication Theory class.

Links as Reinforcers

You should also give careful thought to what pages you want to link to. Sometimes there's a tendency among novice Web writers to link to everything, calling the result "my favorite sites" or "cool pages." Although there can be some charm in eclectic site lists (you never know what you might find), they can also seem irrelevant. What do they have to do with the page that you're currently on? The Web is full of strange, wonderful, and weird sites, but many users want links to be worth the trouble and bandwidth of clicking. When they click on a link, they prefer links that are clearly related to the site they're currently visiting. If they're interested enough in the topic to come to that site in the first place, they might like to find more information at related sites. At least they're more likely to be interested in related sites than they are in sites on unrelated topics that appeal to the site designer.

Used well, links can reinforce your information by supplying additional connections. Once you've caught your readers' interest in your content, your links can help them search further for other relevant information.

Straightforward Linking

As I've mentioned previously, you need to give your readers a clear idea of where they're heading when they click on one of your links. For most readers URLs alone won't mean much; unless they already know what a site is, the URL won't provide much useful information about where the link is going. In fact, even the name of the site won't mean much unless the reader has already visited the site before. If you're going to link to other sites, it helps to include a short description of what the site contains so that the reader can decide whether to make the journey across the Web.

Unlinking the Current Page

One way to show your familiarity with Web conventions is to be sure that you make the link inactive for whatever page the user is currently on. That is, if I'm on the page called "Spring Classes," then the link to the "Spring Classes" page should not be clickable.

It's tempting to simply remove the link for the current page altogether so that the user doesn't have the option to click it, but that can be confusing. Users should see the same navigation options in the same order on every page; in that way they'll become familiar with the way your site is structured.

Instead, if you create a navigation menu with all of your main pages on it, you should make the link inactive for each page within the code for that page by simply removing the `<a href="url"></a>` tags. As an added bonus, when you remove the `<a></a>` tags from the page link, you'll also remove the link colors, which will be a subtle you-are-here indicator on the menu.

PROCESS FOR INSERTING LINKS

Use the following steps to create links for your pages:

1. Make sure each page contains the basic information required for global navigation: a site id, a way home, utilities, and site sections. If you have access to a search utility, place it on every page.

2. Insert absolute links to all off-site pages. In the link text, identify the site being linked to and include a `title` attribute. Include a short description of the site content along with the link.

3. On every page insert a link to a contact e-mail (this can be part of your utilities).

4. Link all the pages of the site together using relative links.

5. If you choose to use images for some links, be sure the images are clear and unmistakable. Insert accompanying text if the images are not self-evident.

6. Insert anchor names or ids along with headings on your longer pages. Create a menu of on-page links at the top of longer pages, and create return-to-top links further down the pages.

7. Create style rules for `a:link`, `a:visited`, `a:hover`, and `a:active`.

CHAPTER SUMMARY

Navigation Basics

❑ Navigation for a Web site will grow out of the organization of the site's information. Navigation represents a way of revealing the site's structure, while showing the user how to move around and how to search for information.

❑ Every page on a Web site should contain the site's global navigation, including site ID, home button, search box, utilities, and major site sections.

❑ Links connect Web resources using source and destination anchors.

❑ URLs are Web page addresses that indicate the location of given pages or files. A typical URL indicates the Web protocol used, the domain of the Web site, and the path to the file.

Absolute and Relative Links

❏ Absolute links include all the parts of the URL address. Relative links include only the part of the URL that changes among pages.

❏ Both absolute and relative links use the `<a href="URL"></a>` format, but the complete URL is used with absolute links, and a shortened version is used with relative links.

❏ Clickable text is placed between the `<a href="URL">` and `</a>` tags. This is the text that will be clicked by the user to reach the Web page in the link.

❏ E-mail links use a slightly different link format: `<a href="mailto:e-mail address"></a>`. All Web sites should include the e-mail address of a contact person for the site.

❏ Images can be used as links. The image file tag is placed between the two link tags: `<a href="URL"><img src="logo.gif"></a>`.

❏ Images have some drawbacks as links: they are more difficult to maintain, and they do not change color once they have been visited.

On-Page Links

❏ On long Web pages, page sections can be linked to a menu at the top of the page, and periodic links can be provided in the body of the page so that users can return to the page top without scrolling.

❏ On-page links either use the `<a name="anchorname"></a>` format or insert an `id` into the tag for the destination link, for example, `<h2 id="Foundations">Foundations of Communication</h2>`.

Attributes

❏ If a `title` attribute is placed within the link tag, a short prose description of the link will appear in a tool-tip box when a cursor is passed over it.

❏ The `target` attribute opens a page in a new window. However, this attribute creates problems for some users and has been deprecated in XHTML.

CSS

❏ CSS can be used to change link styles through four pseudoclasses: `a:link` for unvisited links, `a:visited` for visited links, `a:hover` to create an effect when the cursor is passed over a link, and `a:active` to change styles as a link is clicked.

Link Concerns

❏ Too many links within a text passage can be distracting. Designers may want to limit in-text links, using link lists instead.

❏ Links should be relevant to the subject matter of the site. Well-used links can reinforce the information on a site by supplying additional resources.

- ❏ In-text links should clearly indicate where a link will go when clicked.
- ❏ You can remove the `<a></a>` tags from the link for the current page to create a you-are-here indicator.

HTML/XHTML TAGS IN THIS CHAPTER

TAG	EFFECT
`<a></a>`	Indicates a link

HTML/XHTML ATTRIBUTES IN THIS CHAPTER

ATTRIBUTE	VALUE	EFFECT
`href`	URL	Indicates URL of link
`name`	Name of anchor	Indicates anchor for on-page link
`id`	Name of anchor	Indicates anchor for on-page link
`title`	Text description of link destination	Displays when cursor moves across link

CSS SELECTORS IN THIS CHAPTER

SELECTOR	PROPERTIES AND VALUES	EFFECT
`a:link`	Font and text properties; colors; backgrounds	Sets the style for unvisited links
`a:visited`	Font and text properties; colors; backgrounds	Sets the style for visited links
`a:hover`	Font and text properties; colors; backgrounds	Sets the effects for active links when the cursor passes over them
`a:active`	Font and text properties; colors; backgrounds	Sets the style for links when they are clicked

SOURCES

Krug, S. 2000. *Don't make me think: A commonsense approach to Web usability.* Indianapolis, IN: Que.

Lynch, P. J., and S. Horton. 2001. *Web style guide: Basic design principles for creating Web sites,* 2nd ed. New Haven, CT: Yale University Press.

Nielsen, J. 2000. *Designing Web usability.* Indianapolis, IN: New Riders.

Shea, D., and M. Holzschlag. 2005. *The Zen of CSS design.* Indianapolis, IN: New Riders.

Veen, J. 2001. *The art and science of Web design.* Indianapolis, IN: New Riders.

Web Site Design

DESIGNING

For most designers, actually planning the design for their pages is the most challenging and enjoyable part of creating a Web site. The design for your site should grow out of your site's content and should complement and enhance the information you're presenting. In this chapter we'll cover some basic guidelines for page design, including visual hierarchy, page dimensions, and establishing a consistent look for your pages. We'll also discuss accessibility, a major topic in modern Web design. At the end of the chapter we'll cover some things *not* to do, some Web design errors that most users find annoying. And in chapters 10 and 11, we'll discuss how to create the design you come up with in CSS.

FINDING DESIGN IDEAS

You should begin thinking about your design by looking at other Web sites, particularly those on similar topics, to get ideas. Consider what color combinations seem appropriate for your content. What kinds of images would convey the particular mood you want to convey? What kinds of page designs not only catch your attention but present the page content in a way that makes it both usable and attractive? The more sites you look at, the more ideas you'll probably find for your own design.

You should also talk to other people in your organization and in your potential audience, if possible. Ask them what Web sites they like and why and what things annoy them about site design. Bookmark the sites that seem most interesting to you, and then look for aspects of their designs that might work for you.

One of the advantages of using CSS to design your pages is that you can actually add the page design after you've created your basic HTML page structure. You'll need to look back over what you have created so far. Chances are you've already come up with some color combinations and images. Concentrate now on how you want to put them together.

EXERCISE: LOOK FOR IDEAS

Visit several sites that have designs you like and that have features you think might be appropriate for your own site. Try sketching a rough design for your pages, incorporating some of the features you've found.

PAGE DESIGN BASICS

There are a few design principles that all effective Web pages share: they have a focal point, they use a visual hierarchy, and they work with the dimensions of the page.

Creating a Focal Point

Like print pages, Web pages need a **focal point** to keep the design organized; one element on the page should dominate the others, creating a center of interest. Your focal point should be placed on the first screen readers encounter on a page, which means it will be placed in the top part. The other elements on the page will be subordinate to the focal point, creating a sense of organization among the page elements. If all elements on the page are visually equal in importance, nothing stands out and readers have difficulty knowing where to begin, what to look at first. A focal point lets readers know what the most important element on the page is and creates a context for the other elements.

Notice in Figure 9–1 how the page title, "Plan Your Visit," draws your eye because of its bright red color and oversized type, as well as its position at the top of the page. Thus, the title becomes an effective focal point for the screen, directing your eye to the subheads and giving you a place to begin exploring the page.

Using a Visual Hierarchy

The focal point helps to establish a **visual hierarchy** on your page, another requirement for effective page design. Steve Krug (2000) suggests that your first objective should be to create this clear visual hierarchy on the screen (31). Web page readers are likely to skim information quickly, particularly if they're looking for a particular topic. Establishing a visual hierarchy means making sure the most important information on the page stands out. The page title, for example, should be large and placed in a prominent place, preferably close to the top of the page. Text sections should be broken into segments, and headings should clearly describe the information found beneath them so that readers can scan material quickly. In addition, related material should be grouped together in the same part of the page; elements of global navigation, for example, should be grouped together in one place and should be found in the same location on every page so that users learn where to look for it.

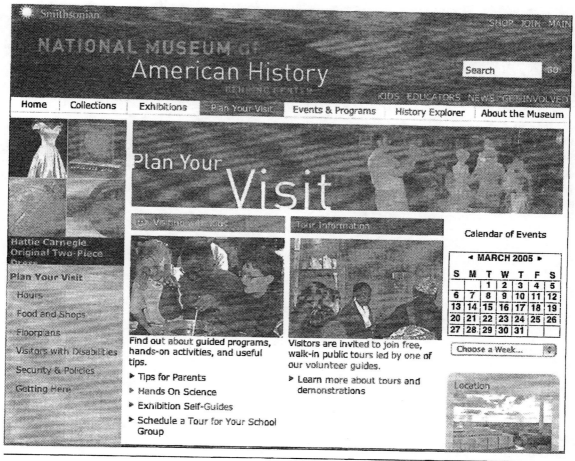

FIGURE 9–1   An example of a focal point

Grouping related material also means that your pages will be divided into clearly defined areas. According to Krug (2000), doing so "allows users to decide quickly which areas of the page to focus on and which areas they can safely ignore." (36). Your visitors may be looking for a particular piece of information and will scan the page to find the areas most likely to have what they need. You can make it easier for them to do that by dividing the page into clearly defined sections, as shown in Figure 9–2.

Notice how the page is clearly divided into navigation on the left (with search at the bottom), page title (and site logo) at the top left, major segments in the center, and utilities across the bottom. You can tell at a glance what parts of the page belong together and what parts are of greatest importance.

FIGURE 9–2   A page showing visual hierarchy

Page Width

Designing the content of your site sections will depend in part on knowing how much space you have to work with. It's important to remember that the screen is smaller than the page. Even though this may seem self-evident, the ramifications of that sizing may not be obvious. First of all, you need to decide how much of the browser window you can use for your page. According to Browser News, by February 2005 around 65 to 70 percent of users had screen resolutions of 1024 × 768 pixels or better. However, a sizable minority of users (up to 28 percent) still relied on 800 × 600 pixels. For these users Web page elements that are wider than 800 pixels can require both horizontal and vertical scrolling, a real annoyance. Thus, given the sizable number of users with smaller screen dimensions, it makes sense to use 800 × 600 as a maximum screen size for your designs. Patrick Lynch and Sarah Horton (2001) argue that the maximum safe size for this screen size is 760 × 410 pixels, and images that must fit onto a page printed from a Web page should be even smaller—560 × 410 pixels.

Should you use pixel measurements to size all of your page elements? Not necessarily. Most designers recommend using relative sizes for elements like text blocks, sizing them using percentages, for example. In that way these elements can change size to fit both higher and lower resolutions, accommodating both ends of the user spectrum.

Page Length

Although your pages can obviously extend beyond 600 pixels in length, there are good reasons not to make them too long. Some Web research has indicated that for most pages, a length of two and a half screens is the maximum that users will tolerate. In fact, in early studies Jakob Nielsen (2000) found that only 10 percent of users would scroll beyond an initial screen (112). Even though that percentage has risen in recent years, the majority of users still prefer to make their navigation selections from those available at the top of the page. In fact, you may find that users miss options that you put below the first 600 pixels of content. Although there are certainly exceptions (e.g., online reports), there are definite advantages to having shorter pages. Readers are more likely to stay with you and are more likely to find what they want if they don't have to search through a long page.

According to Lynch and Horton (2001), the only exception to this advice is material that you don't expect your users to read online (e.g., a technical report). Since the user will most likely be printing the page and reading it off line, it makes sense to include all the content on one page, even though that page extends beyond two and a half browser screens.

Graphic designers Robin Williams and John Tollett (2000) recommend creating an 800 × 600 pixel template by making a screen capture of an 800 × 600 pixel browser window and then placing it in a graphics program. Page designs that you try out in the graphics program can then be placed on top of the browser template to make sure that your designs don't exceed the area you have available.

CONSISTENCY

Once you establish the design you'll use for your pages, you'll need to apply that design consistently throughout your site. Be sure that your navigation always appears in the same place on every page, for example, so that your readers will know where to look for it. And if you have a centered page title on one page, don't switch to a left-justified title on another (unless you're using a different design for another section of your site). In addition, keep the same color scheme at the same level of hierarchy for all your pages. If all the pages in one section of your site have light blue backgrounds for the title section while all the pages in another section have cream-colored backgrounds, your readers will know immediately which section they're in.

Furthermore, although your home page may have a different design to make it distinct from the other pages on your site, it should have some features that will repeat on the other pages. For example, you might use a graphic on your home page and then repeat that graphic (or a portion of it) in a smaller size on your main section pages. If you have a logo for your organization, you can introduce it on your home page and then repeat it on the other pages. And as I've mentioned previously, you should certainly repeat the navigation from your home page on your other pages. The position of the links may be different on the home page and the interior pages, but the link names used on your home page should be repeated on the other pages.

Page Templates

The easiest way to create consistency among your pages is to set up a page **template** that you use for each page on your site, or at least each page at a certain level of hierarchy or in a certain section. A page template ensures that all the pages have the same features, the same layout, and the same color scheme.

The first step in creating your page template is to determine what content you want to include on every page in your site. Remember Steve Krug's list (2000) of navigation elements that you need on every page: a site ID, a way home, a search utility (if you have one), other utilities, and site subsections. Add to these a page title and page content sections, and then try some possible arrangements. The site ID most frequently goes in the upper left corner of your page, and the page title should also be placed at the top of the page. To position the rest of the elements, you'll need to think about the way you want the visual hierarchy to show up.

In chapter 11 we'll discuss several possibilities for column layout. You might decide to use a traditional Web page layout with navigation in a column on the left and content in a column on the right, as in the home page from the Environmental Protection Agency (see Figure 9–3). Or you might decide to have your global navigation at the top of the page and your local navigation (i.e., links to pages within a particular section) on the left or right, as in the page from the National Endowment for the Humanities (see Figure 9–4). You'll also want to consider where you want images; for example, do you want a graphic to appear in the same place on all of your pages, or do you want to insert images at appropriate points in the text and then have the text wrap around them?

Ideally, after you've worked with the elements that will show up on almost every page of your site, you'll be able to come up with a template design that will be clear and consistent, pulling your pages together through a shared design. CSS will make this easier to do by allowing you to use style sheets that link the same design to every page in a particular section on your site. Once you have your design roughed out, you can use chapters 10 and 11 to help you write the CSS that will set up the styles for your pages and then write the HTML for the template pages. After that, you'll simply place the content for each page in the areas you've laid out on the template.

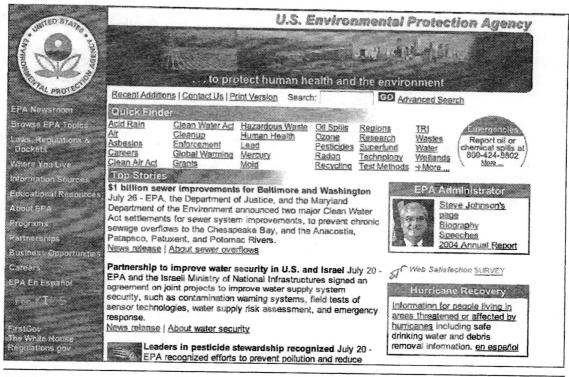

FIGURE 9-3 Layout with a navigation column

EXERCISE: DESIGN YOUR PAGE TEMPLATES

Using the preliminary designs you sketched in the previous exercise, try coming up with a design for each section of your site. You can make these sketches with pencil and paper, or you can use a graphic program like Photoshop or Freehand. Don't worry about coding at this point; just get an idea of the basic page template designs you want to use.

BROWSER CHECKS

As you develop your styles, you should check them frequently to see how they'll look in various browsers. You may have one major browser that you use for the majority of your Web surfing, perhaps another fallback browser that you use if your favorite browser isn't available. However, when you check your pages, you can't confine yourself to just one browser, no matter how popular that browser may be. Even if only a small percentage of your users visit your site with another

FIGURE 9–4 A page with horizontal navigation

browser, you still need to know what they'll see when they visit. You should have recent versions of the most commonly used browsers on your computer, and you should check your pages in all of them.

You'll probably discover that your pages look slightly different from one browser to another. That's not a serious problem; it's impossible to have complete control over your pages so that they look identical in every browser. And keep in mind that your visitors may be using different default settings in their browsers, which will also affect the way your pages show up. Minor differences in spacing or in font size may be simply beyond your control.

However, you should be concerned if your pages look radically different from one browser to another. In particular, if one browser isn't reading your style sheets or if your images don't come up in one browser, you must try to find out what's going on. Some browsers are more strict than others in their requirements for correct HTML and CSS. If one of these browsers isn't showing part of your page or if it's showing up in a radically different way, there's a good chance that you have a problem in your code that needs to be corrected.

ACCESSIBILITY

Accessibility is another factor you must consider as you plan the design of your site. The accessibility movement is concerned with making Web content available to the widest possible audience. On accessible Web sites, users with disabilities can obtain the same information and perform the same tasks as any other users. However, accessibility benefits more than just readers with disabilities. In general, making a site accessible creates a better experience for all those who visit the site.

Making the Web accessible for all users has been a concern of the W3C since 1999, when the first version of the Web Accessibility Initiative was issued. (The most recent version, 2.0, was issued in May 2004.) According to the consortium, "Web accessibility means access to the Web by everyone, regardless of disability" and includes "Web sites and applications, Web browsers and media players, Web authoring tools, and evolving Web technologies."

Paul Bohman (2003) of the Web Accessibility in Mind (WebAIM) organization points out that access to the Web means that visually disabled users can read newspapers on the day they're published, rather than waiting for a Braille printout to be issued several days later; people with motor disabilities can have access to printed information without having to hold the document; and hearing-disabled users can read transcripts of speeches and see captioned versions of multimedia content. However, all of these possibilities are available only if Web designers design them into the pages they produce. Thus, the accessibility of the Web to all users rests on the willingness of designers to make their Web sites accessible.

Studies estimate that around 20 percent of the U.S. population has some kind of disability, although not all of these disabilities make the Web difficult to use. Still, a significant number of individuals with disabilities do have problems accessing Web information. Many public institutions, such as schools and universities, are required by law to make their information available to all users; Section 508 of the Federal Rehabilitation Act contains accessibility requirements for federal Web sites, and many states have instituted their own requirements as well. Internationally, the W3C's Web Accessibility Initiative guidelines have been accepted by members of the European Union, Canada, and Australia.

In addition, many sites that are not required by law to follow accessibility guidelines frequently choose to do so simply because they want to make sure that the largest possible number of visitors can reach their pages. Even those designers who are convinced that few users with disabilities will want access to their sites (and that assumption can usually be challenged) should consider this point: in many cases the adaptations that make sites accessible to users with disabilities also make sites more accessible to those with Web-enabled devices like cell phones and PDAs— a group most Web site owners would be sorry to lose.

Accessibility Guidelines

Fortunately, most of the adaptations needed to make Web sites accessible are part of good markup in the first place. In fact, if you're following the coding guidelines outlined in the previous chapters (using CSS for presentation and HTML for structural markup), you're more than halfway there. Here are some specific recommendations for making your Web pages accessible to all users.

Use Text Equivalents for Images. Users with vision disabilities use text-to-speech synthesizer programs (i.e., text readers) to convert Web content to sound. Obviously text readers can't read images, but Web designers can use the `alt` attribute (described in chapter 6) in their `img` tags to supply a text description for the graphic. These attributes also supply information for text-only browsers, such as those used by some Web-enabled devices. Using `alt` is one of the easiest yet most important accessibility features.

Make Links Clear in Any Context. Text readers identify links by labeling them as they read. However, uninformative link text can create barriers for disabled users and may create confusion for sighted users as well. Thus, rather than "Click here," link text should clearly indicate where the link goes, for example, "College of Liberal and Fine Arts Home." As we noted in the previous chapter, the text that users click on should make the destination of the link clear, and links to the same place should use the same text. Again, following this guideline will not only help disabled users but will also make your pages more usable for everyone.

Use Valid Page Markup. As I've said before, HTML is used to indicate the structure of your page. If you use HTML in this way, text readers and other assistive technologies can easily translate the page into the form needed by a disabled user or a user with a Web-enabled device. On the other hand, if you use HTML in the wrong way (e.g., using a block quote tag simply to indent text rather than to present a quotation), you may make it difficult for users with assistive technology to navigate your pages and understand your content.

Use CSS. CSS is the other part of valid markup; it enables HTML code to be restricted to structural information. Because text readers don't read style sheets, styles that have been placed there won't interfere with the content of the site. Moreover, using style sheets ensures that you use valid page markup, avoiding invalid code (e.g., using tables for page layout rather than for data).

Use Relative Units. Relative units, such as percentages for widths or keywords for font sizes, will resize your content to fit the needs of the user's screen. Thus, your pages will fit on a cell phone screen or will be automatically enlarged to fit on a screen that has been modified for a user with a vision disability. Using fixed units, such as points, can create problems for specialized devices.

Pay Attention to Color and Contrast. Information conveyed by color alone may be lost to users with vision disabilities. Moreover, users with color blindness (around 8 percent of men and 0.5 percent of women) may also be unable to use information that is conveyed by colors they can't see. Thus, if color carries information on your page (e.g., if you indicate which fields of a form must be filled out by making them red), that information should be conveyed in other ways as well (e.g., marking the required fields with an asterisk in addition to the color red). You should also make the contrast between colors as sharp as possible; bright colors are more visible even for those with color blindness.

Most of these guidelines don't require any extraordinary measures. However, they do require designers to work deliberately, being aware of the choices they make and the code they use. Ultimately, working in this way will result in a better experience for *all* users.

BAD DESIGN

Since we've looked at the elements that make page designs effective (i.e., creating a focal point, using a visual hierarchy, and so on), it may also help to look at the things that can make page designs *ineffective*. Even though Web designers and users may disagree about what constitutes the best Web design, they seem to agree about what makes Web design ineffective, if the many lists of Web design mistakes are any indication. These lists may differ in the way that they phrase what the authors consider to be mistakes, but many of the mistakes themselves seem to be consistent from list to list. What follows are design blunders to be avoided, organized into general categories: presentation, navigation, text, and images. We've covered many of these guidelines in earlier chapters, but it may help to see them gathered into one list.

Errors in Presentation

Combinations of Background Colors and Text Colors That Are Hard to Read. Color can enhance your site's message or can work against it. Choose color combinations that have high contrast and that don't cause eye strain. Dark text against a light background is usually the most legible combination. Highly saturated complementary colors can seem to vibrate and cause eye strain.

Background Patterns That Interfere with Text. Like colors, background pictures can be used to make a site attractive, but they should never be so busy that they interfere with text. Be sure to check combinations of text and images to see if they work together.

Splash Pages. Splash pages (i.e., opening pages that are designed to catch the reader's attention before moving on to the home page) were innovative in the midnineties. Today they're a clear indication that a design is dated. In addition, because many readers now enter a site via a search engine, they may never see the opening page at all, thus making it wasted space. Every page should have content, and your opening page should serve as a home page, with at least an introduction to your site and links to your other main pages.

Frames. Frames are another feature that represented an innovation in the early days of the Web but are now outdated. At best, frames create problems if you try to bookmark pages; bookmarks usually list the opening "frameset" page, but not the inner "frame" page. At worst, they fragment the page into chunks separated by borders and take longer to load than the equivalent HTML and CSS.

Browser-Specific Design. It's always a bad idea to use a design that restricts your users to a particular group. You compound the error by adding statements like "Best when viewed with Browser X," which may seem to intrude on a user's browser choice.

Design That Conceals the Site's Subject and Purpose. Vincent Flanders and Dean Peters (2002) argue that "you should be able to look at the home page of any site and figure out what the site is about within four seconds." If your users can't figure out who you are and what your site is about, your design is a failure.

Inconsistent Presentation from Page to Page. Your users are learning to use your site as they move around it and will come to expect a certain look and layout from one page to another. If you arbitrarily change the way your links look or their placement on the page, for example, users are likely to be both confused and annoyed. You should use the same link colors consistently and try to keep your links in the same place on all your pages.

Errors in Navigation

Visited Links That Don't Change Color. Usability studies have shown that users appreciate being able to tell the difference between links they've already clicked and those that they haven't. Visited colors are easy to add in most cases and improve your site's navigation immensely.

Links That Don't Clearly Indicate Link Destination. Telling users to "click here" won't help much unless they know where they're going when they do.

A Link to the Current Page on the Current Page. It's confusing to click on a link and go nowhere; consequently, the page the user is on should not have a live link on the menu. However, retaining the name of the current page on the menu and making the link inactive (i.e., removing the anchor tags) provides a quick and easy you-are-here indicator.

No Links Back to the Home Page. As Steve Krug (2000) points out, home page links are reassuring. They indicate that no matter how deeply users venture into the site, they can always get back to the beginning.

Unmarked and Unclear Icons for Buttons. Vincent Flanders and Dean Peters (2002) refer to this kind of navigation as "mystery meat." Sometimes users can get the definition for the link by passing a cursor across it (i.e., if the designer has included `title` attributes), but how many will bother to do that? Many users will simply head for another site.

Outdated Links That No Longer Work. Links that no longer work are referred to as link rot. All links should be reviewed periodically to make sure the URLs are still correct. A page with several inactive links (i.e., links that produce page-not-found responses) may seem dated and its information suspect.

No Contact Information. All sites need an e-mail address for a contact person. If nothing else, there should be a link to the site designer.

Errors in Images

Large, Slow-Loading Images. Images take time to load, even over the fastest Internet connection. Large images that haven't been compressed in any way will slow down any site. At the very least, graphic tags should include width and height attributes so that browsers can use placeholders for images and download the rest of the page content. At best, however, the designer should make sure that graphic files are kept as small as possible.

Unnecessary or Meaningless Images. Images can help to illustrate text, set the mood of the site, and provide extra information about a topic. But images shouldn't be used just to take up space. If you find yourself wondering why the designer has used that pink flamingo picture, there's a good chance it doesn't belong on the page. There should always be some clear relationship between image and site so that users never feel their time has been wasted waiting for an unnecessary download.

Images without `alt` *Attributes.* As I said earlier, `alt` attributes are one of the simplest and most immediate ways to make a site more accessible. If users cannot see images, either because of vision disabilities or because their browsers do not show images, `alt` tags provide the information that the images convey. This is important for all information-bearing images, but it's vital for linked images, which must have `alt` attributes specifying their link destinations.

Images That Are Too Wide for the Browser Window. For a browser window that is 800 × 600, images should not be wider than 760 pixels. If you use a graphic that exceeds 800 pixels, users whose browsers are set to 800 × 600 will have to scroll both horizontally and vertically, and they may simply skip your site.

Unnecessary Animations. All animations, including animated GIFs, take longer to download than still pictures do. An animation should add something to your page other than glitz to justify including it. Users will not appreciate waiting several seconds for a graphic to download, only to discover it's a cheesy animation that has little to do with the page on which it's placed.

Neglecting to Provide Graphic Files to Accompany HTML Files That Use Images, or Not Checking the Graphic Files in a Browser after Uploading. Missing images will undermine any impression you try to make with your site. They make a designer seem careless at best and unskilled at worst. Make sure that your graphic files are uploaded, and be sure that the files themselves are viewable through a browser (i.e., as GIF, JPEG, or PNG).

"Under Construction" Images. All sites are under construction as conscientious designers update information regularly. Putting "Under Construction" signs on your site merely calls attention to a lack of content and may undermine confidence in the rest of your site, particularly if the signs stay up for too long.

Errors in Text

Text That's Too Small to Be Read Easily. It's hard to say exactly what size text should be (although 9 pixels is probably the downward limit). If your text is too small for users to read comfortably, they may not bother to read it at all. Some fonts are designed to be readable at smaller sizes (e.g., Verdana and Georgia), but even these fonts can't be read if they're too small. At the very least, make small text resizable so that users can enlarge it if they need to. Remember that text in pixels cannot be resized in Internet Explorer.

Too Much Text on a Page. Remember the two-and-a-half-browser-screens rule. If your text runs longer than that at 800 × 600 resolution, consider subdividing your page into multiple pages linked to the main page. The only exception to this rule is a page that your users will most likely print and then read off line. In that situation a long page makes sense, rather than forcing users to open multiple pages and print them out individually.

Text Lines That Stretch All the Way across the Browser Screen. Long text lines are hard to read, and browser windows can be wider than most text pages. Use margins, padding, or relative widths to reduce the size of your text lines.

Solid Text Blocks without Headings or Other Aids to Scanning. Few readers will read every word you write on your Web page. Since most readers will scan your pages, you should make it easier for them to do so by including headings and subheadings, lists, and so on. If users can find what they need quickly, they're more likely to come back to your site—even if they don't read every word you've written there.

Text in ALL CAPS for Emphasis. Text written in all capital letters is difficult to read; we read words, in part, by recognizing their shapes. Consequently, when you print a word in all caps, it makes that shape rectangular and makes it more difficult to recognize the word. If you need to emphasize text, use some other style, such as boldface or italic type. But don't use boldface or italic for long passages; that's also difficult to read.

Centered Text Passages. Although you can use centering for some titles or headings, you shouldn't use centered alignment for paragraphs. Centered text is difficult to read because it's harder to find the beginning of each line as you go from left to right.

Underlined Text That Isn't a Link. In print documents you can use underlining as a form of emphasis. However, on the Web, underlining has become a convention that indicates a hyperlink. If you underline text to emphasize it, users may think that the text is a link and then be perplexed (and perhaps annoyed) when they discover that it isn't. Reserve underlining for links only.

No Clear and Concise Page Title. The caption placed between `<title>` and `</title>` will be used in bookmark and history lists in many browsers. It will also be part of the information listed in search engines. Thus, each page should have a brief title that clearly indicates the page's subject. Moreover, each page title on a site should be distinct. If all your page titles use the same words, it will be difficult to tell which page is which in search results and history lists.

Mistakes in Spelling, Grammar, Punctuation, and Syntax. Not only will these mistakes undermine users' trust in your site, but they may actually make the site difficult to understand and use.

PROCESS FOR DESIGNING PAGE TEMPLATES

Use the following steps to create your page template designs:

1. Study other Web sites on similar topics to find ideas to incorporate into your own design. Keep a list of design features you want to use.
2. Create a rough draft of your page design, either with pencil and paper or with a graphics program. Be sure that the design includes a focal point and an effective visual hierarchy.
3. If you're designing your rough draft in a graphics program, measure it to see how it will look on an 800 × 600 pixel screen.
4. Decide which design elements you'll carry over into other page templates in order to maintain consistency within your site pages.
5. As you begin to transfer your design from the draft to code, work with accessibility guidelines in mind.

CHAPTER SUMMARY

Design Basics

❑ The design of a Web site should grow out of the site's content, enhancing and complementing the information being presented.

❑ Each page should have a focal point around which the design can be organized.

❑ Each page should have a clear visual hierarchy, which allows users to find what they're looking for quickly. Page titles should be prominent, text segments should be clearly divided, and related material should be grouped together.

❑ Pages should be designed with a default minimum size of 800 × 600 pixels.

❑ Page design should be applied consistently so that all pages at a particular level within a site use the same design template.

❑ Consistency can be maintained by designing a set of page templates that can be applied to each level of a site's design.

❑ Designs should always be checked in a variety of browsers, particularly the most recent versions of the major browsers being used.

Accessibility

❑ Web designers should observe accessibility guidelines in order to make their sites available to the widest possible audience, including users with disabilities.

Design Errors

❑ Common errors in presentation include illegible combinations of background colors and/or background images and text, splash pages, frames, browser-specific design, designs that conceal a site's subject and purpose, and inconsistent style.

❑ Common errors in navigation include visited links without a color change, link text that doesn't clearly indicate the link's destination, links to a current page, failure to include a home page link, unmarked or unclear icons for buttons, outdated links, and lack of contact information.

❑ Common errors in images include large, slow-loading images; unnecessary or meaningless images; missing `alt` attributes; images that are too wide for a browser window; unnecessary animations; missing graphic files; and "under construction" images.

❑ Common errors in text include too-small font size, pages with too much text, text lines that extend all the way across the screen, solid text blocks without scanning aids, use of ALL CAPS for emphasis, centered text passages, underlined text that isn't linked, no clear page title, and mistakes in spelling, grammar, punctuation, or syntax.

SOURCES

Bohman, P. 2003. Introduction to Web accessibility (October). http://www. webaim.org/intro/.

Flanders, V., and D. Peters. 2002. *Son of Web pages that suck.* Berkeley, CA: Sybex.

Krug, S. 2000. *Don't make me think: A commonsense approach to Web usability.* Indianapolis, IN: Que.

Lynch, P. J., and S. Horton. 2001. *Web style guide: Basic design principles for creating Web sites,* 2nd ed. New Haven, CT: Yale University Press.

Nielsen, J. 2000. *Designing Web usability.* Indianapolis, IN: New Riders.

Williams, R., and J. Tollett. 2000. *The non-designers' Web book,* 2nd ed. Berkeley, CA: Peachpit Press.

World Wide Web Consortium. 2004. Introduction to Web accessibility. http://www.w3.org/WAI/intro/accessibility.php.

Using CSS for Site Layout: Part 1

DESIGNING SITES WITH CSS

Before CSS was fully implemented in browsers, designers who wanted some control over their page layouts used a combination of invisible layout tables and design hacks like single-pixel GIF spacers to create their page designs. Now, however, CSS positioning enables designers to create basic layouts with much simpler, more-efficient code.

At this point you should have a collection of HTML pages and CSS style sheets from earlier chapters. You may also have a rough draft of some designs for page templates. Fortunately, using CSS to set up these designs won't require you to completely redo what you created earlier. Instead, you'll add some tags to the existing HTML and some rules to the existing CSS.

In this chapter we'll begin with some basic information about CSS layout, discussing the CSS box model and the flow of elements within your page. We'll also talk about the simplest form of CSS layout, the float, which you first encountered in chapter 6.

GENERIC TAGS: DIV AND SPAN

As we move into positioning, you'll encounter two more HTML tags: `<div></div>` and `<span></span>`. Both tags were actually part of earlier HTML specifications but have begun to be used more widely with the development of CSS layout.

SIDEBAR: CSS POSITIONING TERMS

Box model: The CSS box model is a tool for visualizing HTML block-level elements. These elements can be seen as a series of boxes containing other boxes, with properties such as padding, spacing, and borders. The largest box is the page itself, within which block-level elements can be positioned relative to the edges of the page. Elements can also be positioned within other elements on the page (e.g., paragraphs can be positioned within divs).

Containing block: Positioned elements may be placed relative to a containing block. For a `fixed` element, the containing block is the viewport, that is, the browser window. For a relatively positioned element, there is no containing block; the element is placed relative to its normal position in the document flow. For an absolutely positioned element, the containing block is its parent element (i.e., `body` for an absolutely positioned div, but `div` if a paragraph is absolutely positioned within a containing div).

Float: Floating elements are taken out of the normal document flow, and the other elements on the page wrap around them. Images are the most common example of floated elements.

Flow: This describes the order in which elements are placed on a page. If no positioning has been used, the first page element appears at the upper left corner, and subsequent elements are placed either beside or below it, depending on properties such as `width` or `float`.

Div

Div stands for *division;* divs are block-level elements that can contain sections of your page between `<div>` and `</div>`. In fact, it may help to think of a div as a container tag. Divs can contain both text and images, block-level and in-line elements—they can even contain other divs. The only default formatting for a div is the block-level line break. Because they have no built-in styling, divs are frequently used with classes and ids to apply various types of formatting (including positioning).

Span

Divs are generic block-level elements; spans are generic in-line elements. You can use `<span></span>` to apply styles to sections of text within a paragraph or other in-line elements. Like `div,` `span` has no built-in formatting; it has only the styles that you assign with classes and ids.

THE CSS BOX MODEL

Positioning in CSS is based on a **box model,** which is basic to all block-level elements (see Figure 10–1). Every block-level element has its own invisible box, inside which is your content. The content can have a border around its outer edge, with padding between the content and its border. Margins add space outside

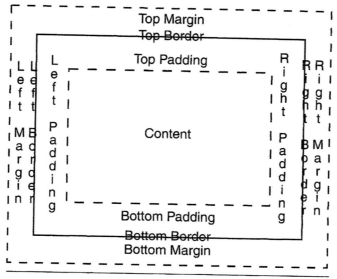

FIGURE 10–1   The CSS box model

the border, putting space between the element inside the box and other elements on the page. The element background (i.e., color, image, or a combination of the two) extends to the outer edge of the border, filling both content and padding as well as gaps in the border if you're using a dotted or dashed border style. On a Web page, the boxes would look like Figure 10–2 (I've put borders around all the block-level elements).

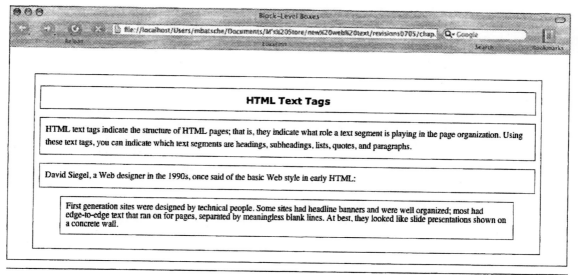

FIGURE 10–2   A page with boxes

Width

In CSS you can set the width for each of these parts of the box—content, padding, border, and margins. You've already seen the measurements for padding, border, and margins; to set the width for the content, you use the `width` property, although there is no content selector. Your selector is the block-level element for which you're setting the content width (e.g., `p`, `div`, `h1`, `blockquote`). The important thing to remember is that these values for `width`, `padding`, `border`, and `margin` are *cumulative*. That is, if you set `width:200px` for a `div`, along with `padding:10px`, and `margin:20px`, the total width of the element box will be 200px + 20px (left and right padding) + 40px (left and right margins), or 260px. When you're trying to position an element, it's important to keep track of how much space that element is really going to take up.

You can use a variety of measurements for `width`, as you could for `padding` and `margin`:

❑ *A length value*—for example, pixels (`px`), inches (`in`), picas (`pc`), or centimeters (`cm`)
❑ *A percentage*—referring to the width of the element's containing block
❑ `auto`—a value that varies depending on the type of element being measured

If the element you're setting the width for has no content, no values will apply because you have no content to measure. Thus, unlike HTML, you can't use empty elements for layout purposes in CSS. In addition, if you have two vertically adjacent elements with margins (e.g., two paragraphs occurring one after another on a page), the bottom margin of the first and the top margin of the next will collapse on each other so that you have only one margin to worry about.

Height

The `height` property is used to set the height for the element box. You've already seen `height` used with images, but it can also be used with other elements, such as divs. As with width, height refers to the height of the element content, and again the values for `height`, `padding`, `border`, and `margin` will be *cumulative*. The same measurements are available for `height` as for `width`:

❑ *A length value*—for example, pixels (`px`), inches (`in`), picas (`pc`), or centimeters (`cm`)
❑ *A percentage*—referring to the height of the element's containing block
❑ `auto`—a value that varies depending on the content of the element

Normally, if an element's height is set to `auto`, the result will be whatever height is necessary to display the content. However, positioning properties such as `top` and `bottom` can override this result.

SIDEBAR: THE BROKEN BOX

Earlier versions of Internet Explorer have problems with the CSS box model. In those versions of Internet Explorer (4 to 5.5 for Windows; prior to 5.0 for Macs), `width` and `height` values are applied to the entire box (except for margins), rather than to the content. In the example cited under `width` in earlier versions of Internet Explorer, the width would be 200px, including the content and the padding. Thus, instead of the 200px-wide content block you were expecting, in these browser versions your content block would be 180px wide (200 − 20px for padding). One way of dealing with this confusion in box models (while still using the correct model for precise layouts) was developed by Microsoft engineer Tantek Çelik. This box-model hack applies a false width that the older browsers can use and then adds a piece of CSS2 code (`voice-family`) that the older browsers won't understand. The older browsers will stop reading the rule at that point, and you can then add the real width for newer, more compliant browsers. The rule looks like this: `#content{width: 220px; voice-family: "\"}\"";  voice-family:inherit; width:200px;}`. Your decision on whether to use this box-model hack should be based on how likely it is that a sizeable segment of your visitors will be using earlier versions of Internet Explorer. At the beginning of 2005, a relatively small number of users still used Internet Explorer 4 for Windows, but a much larger segment used Internet Explorer 5 for Windows.

Flow

Flow refers to the way content flows onto the page as it downloads in CSS. Browsers start placing all content at the upper left corner of the screen and then continue to the right and below as space becomes available (similar to the way content flows in a word processing program).

When you create block-level elements, they're placed in the document flow along with other elements unless you position them elsewhere. If you don't specify the width for a div or a paragraph, the element will span the entire browser window, and the next div or paragraph will begin on a new line since these are block-level elements. Even if you declare a width for your divs or paragraphs, because they're block-level elements, they'll still occur vertically one after the other unless you position them otherwise. This is the default positioning, which is technically called `position:static`. Because it's the default position, you don't have to declare it in a CSS rule, but there are three other types of positioning that you do have to declare: relative, absolute, and fixed. We'll talk more about these types of positioning in chapter 11.

You can actually position some elements without using the `position` property, in effect using `position:static`. For example, if you want to

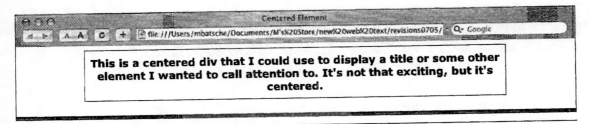

FIGURE 10–3   A centered element

center a div, you can set a width that's less than the width of the browser window, say 75 percent. Then you can set the left and right margins at auto. Using auto as the value for margin-left and margin-right sets the two margins to equal values. And since you've already set the content and padding width, the remaining space will be equally divided between the two margins. Here's a style rule for a centered box (Owen Briggs, 2002):

```
#title{padding:10px; margin-top:5px; margin-bottom:
    5px; margin-right:auto; margin-left:auto; back-
    ground-color:#fff; width:75%;}
```

In a browser it would look like Figure 10–3 (I've added a border):

One problem with this method of using auto, however, is that versions of Internet Explorer for Windows prior to 6.0 don't recognize the value. Instead, you could add another rule to your stylesheet: body{text-align: center}. Even though it's not correct CSS, earlier versions of Internet Explorer will do the kind of centering you're looking for. However, you'll have to override the text-align:center in other CSS rules on the page to return your text to its normal alignment. Most of your layout will be done with floated or positioned elements rather than elements using position: static.

FLOAT

You've already seen the float property used to float images so that the text wraps around them, but it can be used to float text as well. With this type of **float**, an element moves to the left or right edge of its parent element (i.e., its containing block) until it encounters the margin, padding, or border of another block-level element (floated elements automatically become block-level elements). float has two possible values (along with the default none): left and right. An element with the property float:left will move to the left edge of its containing block; an element with float:right will move to the right edge. You should always declare a width with a floated element; otherwise, the results may be unpredictable.

You can easily use `float` to position sidebars and other small elements just as you used it to position images. The CSS looks like this:

```
#sidebar{float:right; width:20%; padding:10px;
    margin-left:20px}
```

The HTML code for a typical page would look like this:

```
<!DOCTYPE html PUBLIC "-//W3C//DTD XHTML 1.0 Transi-
    tional//EN" "http://www.w3.org/TR/xhtml1/DTD/
    xhtml1-transitional.dtd">
<html xmlns="http://www.w3.org/1999/xhtml">
<link rel="stylesheet" href="float.css"
    type="text/css" />
<head>
    <title>Floated Sidebar</title>
</head>
<body>
<div id="sidebar">
<p>
This is my sidebar. It will float on the right side
of the page alongside my text.</p>
</div>
<p>
You've already seen the float property used to float
images so that the text wraps around them, but float
can be used to float text as well. If an element is
floated, it will move to the left or right edge of
its parent element (i.e., its containing block)
until it encounters the margin, padding, or border
of another block-level element (floated elements
automatically become block-level elements). Float
has two possible values (along with the default
none): left and right. An element with the property
float:left will move to the left edge of its con-
taining block; an element with float:right will move
to the right edge. You should always declare a width
with a floated element; otherwise, the results may
be unpredictable.
</p>
<p>
You can easily use float to position sidebars and other
small elements just as you used it to position images.
</p>
</body>
</html>
```

And in a browser it would look like Figure 10–4.

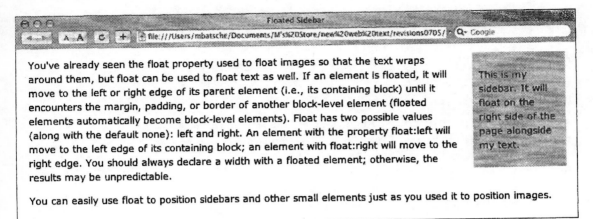

FIGURE 10–4   A simple float

Two-Column Pages

float can also be used to create two-column pages. In one variation one column is floated, and the other is given a wide margin so that it will not wrap around the floating column. The CSS looks like this:

```
#right{float:right; width:40%; padding:10px}
#content{margin-right:45%}
```

Notice that the content column has been given a margin that's 5 percent wider than the width of the floating column. That will provide a generous margin between the two. Typical HTML code would look like this:

```
<!DOCTYPE html PUBLIC "-//W3C//DTD XHTML 1.0 Transiti-
    sonal//EN" "http://www.w3.org/TR/xhtml1/DTD/xhtml1-
    transitional.dtd">
<html xmlns="http://www.w3.org/1999/xhtml">
<head>
    <title>Floating Column</title>
    <link rel="stylesheet" href="float.css" type=
    "text/css"/>
</head>
<body>
<div id="right">
<p>
Float can also be used to create two-column pages.
In one variation, used here, one column is floated,
and the other is given a wide margin so that it
will not wrap around the floating column. Notice
```

```
that the content column has been given a margin
that's 5 percent wider than the width of the float-
ing column. That will provide a generous margin
between the two.
</p>
</div>
<div id="content">
<p>
You've already seen the float property used to float
images so that the text wraps around them, but float
can be used to float text as well. If an element is
floated, it will move to the left or right edge of its
parent element (i.e., its containing block) until it
encounters the margin, padding, or border of another
block-level element (floated elements automatically
become block-level elements). Float has two possible
values (along with the default none): left and right.
An element with the property float:left will move to
the left edge of its containing block; an element with
float:right will move to the right edge. You should
always declare a width with a floated element;
otherwise, the results may be unpredictable.
</p>
</div>
</body>
</html>
```

Notice here that the floated column is placed before the content column in the
HTML. In a browser it would look like Figure 10–5.

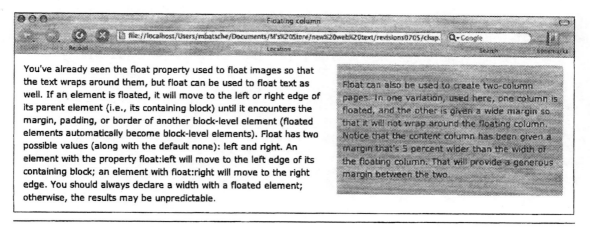

FIGURE 10–5   A two-column page

You can also float both columns in a two-column layout. As Dan Cederholm (2004) points out, this variation has the advantage of placing the content column first in the HTML, which means it will be read first by screen readers and text browsers and in older browsers that don't support CSS. However, the total width of the two columns, along with any padding and margins, **must be less than 100 percent** so that the columns float side by side. The CSS looks like this:

```
#leftcolumn{float:left; width:45%}
#rightcolumn{float:right; width:45%}
```

The HTML would look like this:

```
<!DOCTYPE html PUBLIC "-//W3C//DTD XHTML 1.0
    Transitional//EN" "http://www.w3.org/TR/xhtml1/
    DTD/xhtml1-transitional.dtd">
<html xmlns="http://www.w3.org/1999/xhtml">
<head>
    <title>Floating Column</title>
    <link rel="stylesheet" href="float.css" type=
    "text/css" />
</head>
<body>
<div id="leftcolumn">
<p>
You've already seen the float property used to float
images so that the text wraps around them, but float
can be used to float text as well. If an element is
floated, it will move to the left or right edge of
its parent element (i.e., its containing block)
until it encounters the margin, padding, or border
of another block-level element (floated elements
automatically become block-level elements). Float has
two possible values (along with the default none):
left and right. An element with the property
float:left will move to the left edge of its
containing block; an element with float: right will
move to the right edge. You should always declare
a width with a floated element; otherwise, the
results may be unpredictable.
</p>
</div>
<div id="rightcolumn">
<p>
You can also float both columns in a two-column
layout. As Dan Cederholm (2004) points out, this
```

```
variation has the advantage of placing the content
column first in the HTML, which means it will be
read first by screen readers and text browsers
and in older browsers that don't support CSS.
However, the total width of the two columns,
along with any padding and margins, must be less
than 100 percent so that the columns float side
by side.
</p>
</div>
</body>
</html>
```

In a browser it would look like Figure 10–6.

Floats with Footers and Headers

As you'll see in chapter 11, it's very difficult to stretch a bottom box across an
absolutely positioned layout, but it's relatively easy to do so with a floated layout
like the one just described. We could use either of the two-column layouts we've
covered, but let's take the one in which both columns are floated and add
a footer box like this:

```
#leftcolumn{float:left; width:45%}
#rightcolumn{float:right; width:45%}
#footer{clear:both; padding:20px;}
```

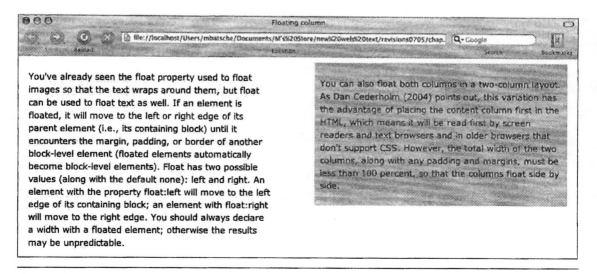

FIGURE 10–6   Two floated columns

The footer uses the `clear` property described in chapter 6, which keeps the float from influencing any of the content that comes after it. Thus, the footer div will stretch all the way across the page, and the HTML will look like this:

```
<!DOCTYPE html PUBLIC "-//W3C//DTD XHTML 1.0
    Transitional//EN" "http://www.w3.org/TR/xhtml1/
    DTD/xhtml1-transitional.dtd">
<html xmlns="http://www.w3.org/1999/xhtml">
<head>
    <title>Footer Column</title>
    <link rel="stylesheet" href="float.css" type=
    "text/css"/>
</head>
<body>
<div id="leftcolumn">
<p>
You've already seen the float property used to float
images so that the text wraps around them, but float
can be used to float text as well. If an element is
floated, it will move to the left or right edge of
its parent element (i.e., its containing block)
until it encounters the margin, padding, or border
of another block-level element (floated elements
automatically become block-level elements). Float has
two possible values (along with the default none):
left and right.
</p>
</div>
<div id="rightcolumn">
<p>
As you'll see in chapter 11, it's very difficult to
stretch a bottom box across an absolutely positioned
layout, but it's relatively easy to do so with a
floated layout like the one just described. We could
use either of the two-column layouts we've covered,
but let's take the one in which both columns are
floated and add a footer box.
</p>
</div>
<div id="footer">
<p>
The footer uses the clear property described in
chapter 6, which keeps the float from influencing
any of the content that comes after it. Thus,
```

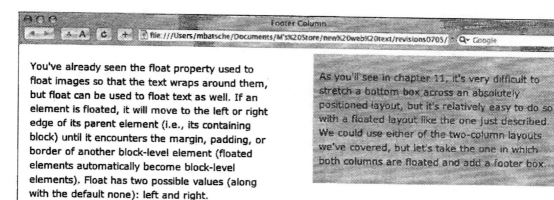

You've already seen the float property used to float images so that the text wraps around them, but float can be used to float text as well. If an element is floated, it will move to the left or right edge of its parent element (i.e., its containing block) until it encounters the margin, padding, or border of another block-level element (floated elements automatically become block-level elements). Float has two possible values (along with the default none): left and right.

As you'll see in chapter 11, it's very difficult to stretch a bottom box across an absolutely positioned layout, but it's relatively easy to do so with a floated layout like the one just described. We could use either of the two-column layouts we've covered, but let's take the one in which both columns are floated and add a footer box.

The footer uses the clear property described in chapter 6, which keeps the float from influencing any of the content that comes after it. Thus, the footer div will stretch all the way across the page.

FIGURE 10–7   A two-column float with footer

```
the footer div will stretch all the way across
the page.
</p>
</div>
</body>
</html>
```

In a browser it would look like Figure 10–7.

If I wanted to have a banner stretching across the top of the page, I would need to place it in the HTML before the two floating divs. Remember that the default width of a block-level element will stretch across the entire page. Thus, a header placed at the top of the page will naturally stretch all the way across. The HTML would look like this:

```
<!DOCTYPE html PUBLIC "-//W3C//DTD XHTML 1.0 Transi-
    tional//EN" "http://www.w3.org/TR/xhtml1/DTD/
    xhtml1-transitional.dtd">
<html xmlns="http://www.w3.org/1999/xhtml">
<head>
    <meta http-equiv="content-type" content= "text/
    html; charset=utf-8"/>
    <title>Float with Footer and Header</title>
    <link rel="stylesheet" href="float.css"
    type="text/css"/>
</head>
```

```
<body>
<div id="header">
<p>
If I wanted to have a banner stretching across the
top of the page, I would need to place it in the HTML
before the two floating divs. Remember that the
default width of a block-level element will stretch
across the entire page. Thus, a header placed at the
top of the page will naturally stretch all the way
across.
</p>
</div>
<div id="leftcolumn">
<p>
You've already seen the float property used to float
images so that the text wraps around them, but float
can be used to float text as well. If an element is
floated, it will move to the left or right edge of its
parent element (i.e., its containing block) until it
encounters the margin, padding, or border of another
block-level element (floated elements automatically
become block-level elements). Float has two possible
values (along with the default none): left and right.
</p>
</div>
<div id="rightcolumn">
<p>
As you'll see in chapter 11, it's very difficult to
stretch a bottom box across an absolutely positioned
layout, but it's relatively easy to do so with a
floated layout like the one just described. We could
use either of the two-column layouts we've covered,
but let's take the one in which both columns are
floated and add a footer box.
</p>
</div>
<div id="footer">
<p>
The footer uses the clear property described in
chapter 6, which keeps the float from influencing any
of the content that comes after it. Thus, the footer
div will stretch all the way across the page.
</p>
</div>
</body>
</html>
```

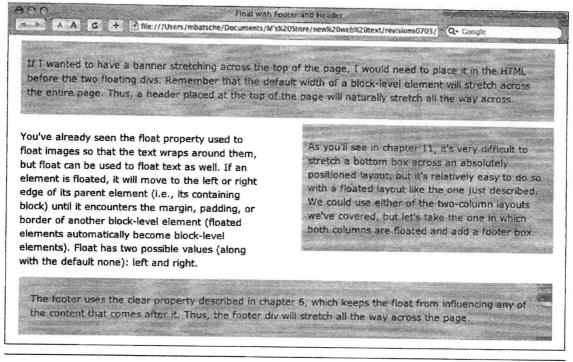

If I wanted to have a banner stretching across the top of the page, I would need to place it in the HTML before the two floating divs. Remember that the default width of a block-level element will stretch across the entire page. Thus, a header placed at the top of the page will naturally stretch all the way across.

You've already seen the float property used to float images so that the text wraps around them, but float can be used to float text as well. If an element is floated, it will move to the left or right edge of its parent element (i.e., its containing block) until it encounters the margin, padding, or border of another block-level element (floated elements automatically become block-level elements). Float has two possible values (along with the default none): left and right.

As you'll see in chapter 11, it's very difficult to stretch a bottom box across an absolutely positioned layout, but it's relatively easy to do so with a floated layout like the one just described. We could use either of the two-column layouts we've covered, but let's take the one in which both columns are floated and add a footer box.

The footer uses the clear property described in chapter 6, which keeps the float from influencing any of the content that comes after it. Thus, the footer div will stretch all the way across the page.

FIGURE 10–8   A two-column float with banner and footer

In a browser it would look like Figure 10–8.

EXERCISE: USING FLOATS

If any of your pages use two-column layouts or footers, try using a float for at least one column. Remember to keep the total width of the columns plus padding and margins to less than 100 percent.

Broken Floats

One problem with using the float property for layout is that floating elements are not fixed in position. Thus, if their total width exceeds the width of the browser window, the floated element on the right will be pushed down below the element on the left. For this reason floating layouts can break, or fall apart, more easily than positioned layouts. For example, if I increased the padding in the two-column example in Figure 10–6 to 30px for both left and right columns, Figure 10–9 would be the result.

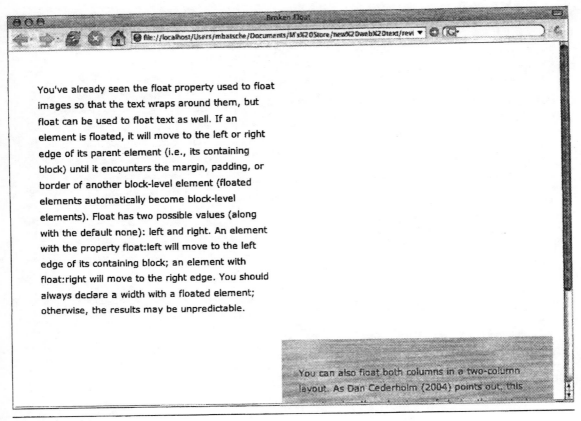

FIGURE 10–9   A broken float

Floats can work for simple, two-column layouts, and they can work more effectively than absolute positioning when you want a bottom element to spread across the page. But for more complex, multicolumn layouts, you'll need to use some form of positioning, which we'll cover in chapter 11.

PROCESS FOR CREATING A TWO-COLUMN LAYOUT WITH **FLOAT**

Use the following steps to create a two-column page layout using the `float` property:

1. Decide which element(s) in your layout will use `float` and which will be stationary.
2. On your CSS style sheet, create an id for the floated element and a style rule that includes (but isn't limited to) the following properties: `float`, `width`, `padding` or `margin`.

3. On your CSS style sheet, create an id for the nonfloated element(s) and a style rule that includes a margin wide enough to move the nonfloated element away from the floated element. For example, if the floated element has a width of 30 percent, the margin for the nonfloated element will be around 35 percent.

4. In your HTML, create divs for the floated and nonfloated elements, using the ids you created on your style sheet.

5. If you want to include a footer on your page, create an id that uses the clear:both property and value, and include any other styles you want for your footer.

6. If you want to create a banner at the top of your page, place a div with the banner styles at the beginning of your body section before the floated and nonfloated elements.

CHAPTER SUMMARY

CSS Positioning Basics

❑ CSS positioning allows designers to create layouts with simpler, more efficient code than older table-based layouts allowed.

CSS Box Model

❑ CSS positioning is based on the CSS box model: every block-level element has its own box, which governs the placement of content, padding, borders, and margins.

❑ Using CSS, you can set the width for each part of the block-level element box. The width property sets the width of the content. Values for width, padding, border, and margin for a single element are cumulative.

❑ CSS also allows you to set the height for an element, although most elements expand the height to fit the content. As with width, the values for height, padding, border, and margin, are cumulative.

Document Flow

❑ Content flows onto the page, starting at the upper left corner of the screen and then moving to the right and down as space becomes available.

❑ Block-level element boxes are placed in the document flow unless you position them outside it. Unless width is specified, a block-level element will stretch all the way across the browser window.

❑ position:static is the default position, placing an element where it would normally be placed in the document flow.

❑ You can center a block-level element without positioning it by declaring a width for the element and then setting the left and right margins to auto. However, Internet Explorer versions prior to 6.0 do not recognize the auto value.

Floats

❑ The float property can be used for layout as well as for an image. Floated elements move to the right or left until they encounter margins, padding, or another block-level element.

❑ A floated element can either have text wrap around it or float in its own space. Floats are more effective than positioning for adding a footer to a page.

❑ Floating layouts are not fixed in position. Thus, slight changes in spacing or width can result in a broken layout.

HTML/XHTML TAGS IN THIS CHAPTER

TAG	EFFECT
`<div></div>`	Block-level container tag without predefined style that can be used for layout
`<span></span>`	In-line container tag without predefined style that can be used for layout

CSS PROPERTIES IN THIS CHAPTER

PROPERTY	VALUES	EFFECT
width	Length value, percentage, auto	Sets width of content within block-level element
height	Length value, percentage, auto	Sets height of content within block-level element
position	static	Positions an element where it would normally be placed in the document flow (default position)
float	left, right, none	Moves an element to the left or right edge of its parent element until it encounters the margin, padding, or border of another block-level element

SOURCES

Briggs, O. 2002. Boxes, boxes, boxes. Chap. 8 in *Cascading Style Sheets: Separating content from presentation*. Birmingham, UK: Glasshaus.

Cederholm, D. 2004. *Web standards solutions: The markup and style handbook*. Berkeley, CA: Friends of Ed/Apress.

Using CSS for Site Layout: Part 2

Chapter
11

POSITIONING

As you saw in the previous chapter, CSS allows you to place information about layout on your style sheet rather than within your HTML. The `float` property, which was introduced in chapter 6 and expanded in chapter 10, works effectively for simple, two-column layouts. However, many pages require more-complex structures than a two-column float, and that's what this chapter is about. We'll cover the three types of CSS positioning: `absolute`, `fixed`, and `relative`; a fourth type, `static`, is the default position and was covered in chapter 10. (All four types are described in the sidebar for easy reference.) We'll also discuss the z-index, which allows you to stack elements on your page. And finally, we'll look at two ways to position navigation menus using CSS.

ABSOLUTE POSITIONING

The `position` property determines whether elements are placed in the normal flow of the page or taken out of the flow and positioned in some other way. Much of your positioning will be done with `position:absolute`. **Absolute positioning** removes elements from the normal document flow and positions them in a particular place on the page. An absolutely positioned element is positioned in relation to its containing block, that is, its parent element. However, the other elements on the page behave as if the absolutely positioned element isn't there, which means you'll have to adjust the position of other elements to keep them from flowing behind the absolutely positioned element.

SIDEBAR: CSS POSITIONING TERMS

Absolute positioning: An absolutely positioned element is removed from the normal flow of content and positioned in relation to its containing block. Other elements behave as if it isn't there.

Fixed positioning: This is a type of absolute positioning. An element that has a fixed position remains on screen at all times. The positioning properties set the distance between the element and the edges of the viewport, which is its containing block.

Relative positioning: Elements that are relatively positioned are taken out of the normal document flow and offset from the sides of the page by amounts declared in the style rule. The position that the element would normally occupy is preserved in the document flow.

Static positioning: With this default positioning, elements remain in their usual places in the document flow. The first element is placed at the upper left, and the other elements are placed beside or below it in the order in which they occur on the page.

The issue of the containing block for an absolutely positioned element can be important. According to the CSS specification, the containing block for an absolute element is the nearest positioned ancestral element (i.e., an ancestral element with a position other than `static`). If there isn't any positioned ancestral element, the containing block is defined as the root element of the page, that is, `<html>` rather than `<body>`. Thus, if you set margins or padding using `body` as the selector, they'll have no effect on the absolutely positioned element. In fact, many CSS designers argue that margins and padding should both be set to 0 for the body, so that spacing takes place entirely with the elements, both positioned and unpositioned. This approach helps to ensure that spacing is exact and also helps you keep track of your spacing so that you don't end up with spaces you didn't expect to have.

Placement Properties

Absolute positioning requires the use of two other properties to give coordinates for the placement of the element: `top`, `right`, `bottom`, or `left`. These properties can have the following values:

- ❏ *A length value*—such as pixels (`px`), inches (`in`), picas (`pc`), centimeters (`cm`).
- ❏ *A percentage*—figured as a percentage of the containing block
- ❏ `auto`

These are **inset** values; that is, the value is the distance *away from* the top, right, bottom, or left edge of the containing block. Thus, if you used `top:15px;`

`left:100px`; the element would be moved down fifteen pixels from the top edge and in 100 pixels from the left edge. Usually, you need to specify only two properties: `top` or `bottom` and `left` or `right`.

Two-Column Layout

As with the `float` property, one of the most common uses for absolute positioning is to create a multicolumn layout on your Web pages. This is the kind of layout that was previously done with tables, using `rowspan` and `colspan`. Now it can be done much more efficiently (and with much less code) using CSS. You can find out more about such layouts in "Boxes, Boxes, Boxes," a chapter in Owen Briggs's *Cascading Style Sheets: Separating Content from Presentation* (2002) or on Briggs's Web site: `http://www.thenoodleincident.com/`. One point should be emphasized from the beginning, however: **not everything on the page needs to be positioned.** Column layout uses a combination of margins, positioning, and document flow to make the columns line up side by side.

In the standard absolutely positioned two-column layout, the menu box is positioned at the left side of the page, with the content box flowing beside it without positioning. With this layout you must declare a width for the absolutely positioned menu because the content is separated from the menu by an extra-large margin that creates space along the side for the menu. The positioning properties on the style sheet would look like this (for a real page you'd also add other properties like font specifications and colors for both divs):

```
body {margin:0px; padding:0px;}
#menu {width:10%; position:absolute; top:10px;
    left:0px; padding:10px 1% 10px 1%; margin:5px 1%
    5px 1%;}
#content {margin:5px 5px 5px 12%; padding:5px 2%
    5px 2%;}
```

The page code would look like this:

```
<!DOCTYPE html PUBLIC "-//W3C//DTD XHTML 1.0
    Transitional//EN" "http://www.w3.org/TR/xhtml1/
    DTD/xhtml1-transitional.dtd">
<html xmlns="http://www.w3.org/1999/xhtml">
<head>
    <title>2-Column Layout</title>
<link rel="stylesheet" href="2-column2.css"
    type="text/css"/>
</head>
<body>
<div id="menu">
<h2>Link list</h2>
<a href="page1.html">Menu item</a><br />
```

```
<a href="page2.html">Menu item</a><br />
<a href="page3.html">Menu item</a><br />
</div>
<div id="content">
<p>
As with the float property, one of the most common uses
for absolute positioning is to create a multicolumn
layout on your Web pages. This is the kind of layout
that was previously done with tables, using rowspan
and colspan. Now it can be done much more efficiently
(and with much less code) using CSS. You can find
out more about such layouts in "Boxes, Boxes, Boxes,"
a chapter in Owen Briggs's <cite> Cascading Style
Sheets: Separating Content from Presentation
</cite> (2002) or on Briggs's Web site:
http://www.thenoodleincident.com/. One point should be
emphasized from the beginning, however: not everything
on the page needs to be positioned. Column layout uses
a combination of margins, positioning, and document
flow to make the columns line up side by side.
</p>
</div>
</body>
</html>
```

In a browser it would look like Figure 11–1.

I used a percentage for the width of the menu and the left margin of the content so that the menu will expand or contract with the size of the window. You could also specify the width in pixels; in that case you could set the left margin on

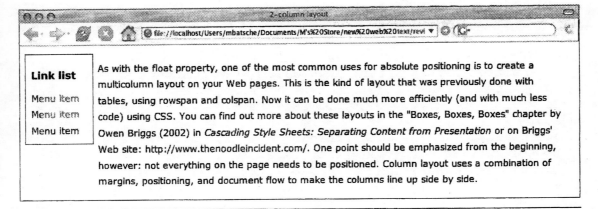

FIGURE 11–1   A two-column absolute layout

the content div to the same number of pixels you used for the width on the menu div. A pixel width will never be too small in the browser window, but it may look too thin on larger screens. Percentages may have the opposite problem, becoming too small on smaller screens. Whether you use percentages or pixels will ultimately depend on your design and your typical users. However, remember that one accessibility guideline is to use relative units like percentages whenever possible.

Three-Column Layout

Three-column design is a mainstay of many Web sites: a menu flows down one side (frequently the left), with other links, news items, or even advertising in the other side column. Then the content flows into a wider box in the center of the page. For this layout you absolutely position the two side columns and then flow the content in between them with wide side margins. The style sheet would look like this, showing only the positioning properties:

```
body {margin:0px; padding:0px;}
#left {position:absolute; top:0px; left:0px;
     padding:10px; margin:5px; width:90px;}
#middle {margin:5px 122px 5px 122px; padding:10px;}
#right {position:absolute; top:0px; right:0px;
     padding:10px; margin:5px; width:90px;}
```

The page code would look like this:

```
<!DOCTYPE html PUBLIC "-//W3C//DTD XHTML 1.0
     Transitional//EN" "http://www.w3.org/TR/ xhtml1/
     DTD/xhtml1-transitional.dtd">
<html xmlns="http://www.w3.org/1999/xhtml">
<head>
<title>3-Column Layout</title>
<link rel="stylesheet" href="3-column.css"
     type="text/css"/>
</head>
<body>
<div id="left">
<p>
Three-column design is a mainstay of many Web sites:
a menu flows down one side (frequently the left),
with other links, news items, or even advertising in
the other side column.
</p>
</div>
<div id="middle">
<p>
As with the float property, one of the most common
uses for absolute positioning is to create a
```

multicolumn layout on your Web pages. This is the kind
of layout that was previously done with tables, using
rowspan and colspan. Now it can be done much more
efficiently (and with much less code) using CSS. You
can find out more about such layouts in "Boxes, Boxes,
Boxes," a chapter in Owen Briggs's <cite>Cascading
Style Sheets: Separating Content from
Presentation</cite> (2002) or on Briggs's Web site:
http://www.thenoodleincident.com/. One point should be
emphasized from the beginning, however: not everything
on the page needs to be positioned. Column layout uses
a combination of margins, positioning, and document
flow to make the columns line up side by side.
</p>
</div>
<div id="right">
<p>
The content flows into a wider box in the center of
the page. For this layout you absolutely position the
two side columns and then flow the content in between
them with wide side margins.
</p>
</div>
</body>
</html>

And in a browser it would look like Figure 11–2.

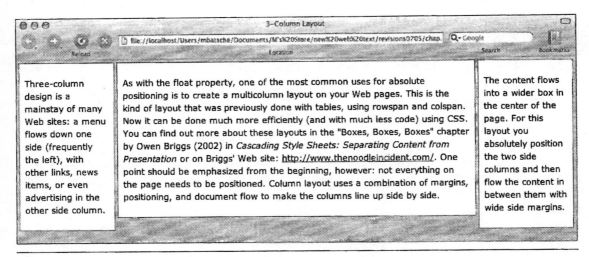

FIGURE 11–2 A 3-column absolute layout

Notice that the three columns here have unequal lengths; that's because the length of a column is determined by the amount of content in it. Setting the height for a column is not recommended. If you set an absolute height in pixels, it may not be tall enough for the content if the browser window size is reduced. And if you use a relative height (a percentage), the percentage will refer to the height of the containing block, which in this case is the page. To calculate height based on a containing block, that containing block must have an explicitly stated height, but you wouldn't want to set a height for your page because you want the page to expand with your content.

Thus, it's not really possible to make the three columns end in the same place, but that's not a really serious problem. Nonetheless, Dan Cederholm (2004) does have a technique for making columns line up equally by using a background image. For more information see Cederholm's book *Web Standards Solutions*.

Three Columns with a Top Row

You can add a fourth box to your three-column layout if you want a section across the top of the page for a title or a logo or other necessary information. To do this, you'll be adjusting the top margin of the content box and the top position of the side boxes. The top position of the side boxes will be the total of the new top box's height, padding, border, and margin. The middle content box will be placed below the top box by the natural flow of the document. You'll use the height property to set the `height` for the top box.

A sample CSS for a three-column page with a banner across the top would look like this:

```
body {margin:0px; padding:0px;}
#top {margin:5px; padding:10px; height:50px;}
#left {position:absolute; top:70px; left:0px;
    padding:10px; margin:10px; width:100px;}
#middle {margin:0px 125px 5px 125px; padding:10px;}
#right {position:absolute; top:70px; right:0px;
    padding:10px; margin:10px; width:100px;}
```

The page code would look like this:

```
<!DOCTYPE html PUBLIC "-//W3C//DTD XHTML 1.0
    Transitional//EN" "http://www.w3.org/TR/xhtml1/
    DTD/xhtml1-transitional.dtd">
<html xmlns="http://www.w3.org/1999/xhtml">
<head>
    <title>3-Column Layout with Banner</title>
```

```
<link rel="stylesheet" href="3-column2.css"
    type="text/css"/>
</head>
<body>
<div id="top">
Three-Column Layout with Banner
</div>
<div id="left">
<p>
Three-column design is a mainstay of many Web sites:
a menu flows down one side (frequently the left),
with other links, news items, or even advertising in
the other side column.
</p>
</div>
<div id="middle">
<p>
As with the float property, one of the most common uses
for absolute positioning is to create a multicolumn
layout on your Web pages. This is the kind of layout
that was previously done with tables, using rowspan and
colspan. Now it can be done much more efficiently (and
with much less code) using CSS. You can find out more
about such layouts in "Boxes, Boxes, Boxes," a chapter
in Owen Briggs's <cite>Cascading Style Sheets:
Separating Content from Presentation</cite> (2002) or
on Briggs's Web site: http://www.thenoodleincident.com/.
One point should be emphasized from the beginning,
however: not everything on the page needs to be
positioned. Column layout uses a combination of
margins, positioning, and document flow to make the
columns line up side by side.
</p>
</div>
<div id="right">
<p>
The content flows into a wider box in the center of
the page. For this layout you absolutely position the
two side columns and then flow the content in between
them with wide side margins.
</p>
</div>
</body>
</html>
```

And in a browser it would look like Figure 11–3.

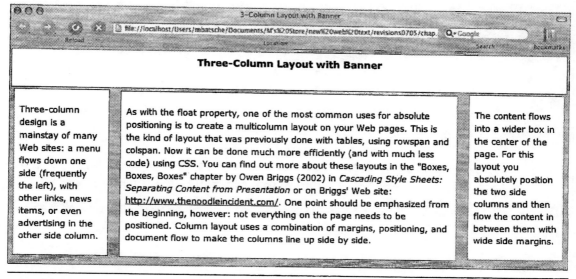

Three-Column Layout with Banner

Three-column design is a mainstay of many Web sites: a menu flows down one side (frequently the left), with other links, news items, or even advertising in the other side column.

As with the float property, one of the most common uses for absolute positioning is to create a multicolumn layout on your Web pages. This is the kind of layout that was previously done with tables, using rowspan and colspan. Now it can be done much more efficiently (and with much less code) using CSS. You can find out more about these layouts in the "Boxes, Boxes, Boxes" chapter by Owen Briggs (2002) in *Cascading Style Sheets: Separating Content from Presentation* or on Briggs' Web site: http://www.thenoodleincident.com/. One point should be emphasized from the beginning, however: not everything on the page needs to be positioned. Column layout uses a combination of margins, positioning, and document flow to make the columns line up side by side.

The content flows into a wider box in the center of the page. For this layout you absolutely position the two side columns and then flow the content in between them with wide side margins.

FIGURE 11–3 A 3-column absolute layout with banner

EXERCISE: USING ABSOLUTE POSITIONING

If any of your page designs involve multicolumn layout, try setting them up with absolute positioning. Remember to use a combination of positioning, margins, and placement to create your pages.

FIXED POSITIONING

Fixed positioning is actually a type of absolute positioning. The containing block for the fixed element is the browser window, which means that an element that uses fixed position will always stay on the screen. If there is an element you want always to be visible (e.g., a menu or a logo), you could place it in a div with `position:fixed`. However, as with other absolutely positioned elements, other elements on the screen will ignore the fixed element, flowing under it as if it isn't there. So unless you want your divs to overlap, you'll need to adjust the spacing.

Here's a style sheet with a fixed div for a logo in the upper left corner and a wide left margin for the rest of the content so that the logo won't overlap the text:

```
#logo {position:fixed; left:0px; top:0px;}
#content {margin-left:100px;}
```

Figure 11–4 shows what it would look like at the top of the page, and Figure 11–5 shows what it would look like as the page is scrolled down.

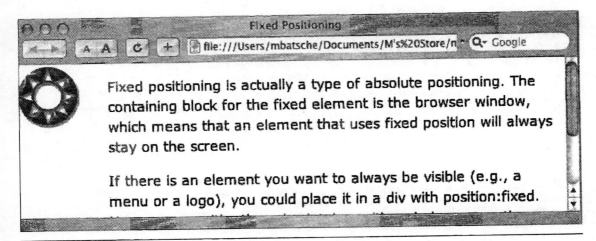

FIGURE 11–4   An example of fixed positioning

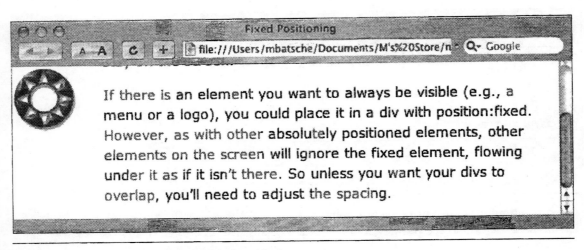

FIGURE 11–5   Another example of fixed positioning

EXERCISE: FIXED POSITIONING

If any of your page designs have elements that you want to remain on-screen at all times, try setting them up with fixed positioning. Remember to use a combination of positioning, margins, and placement to create your pages.

Dave Shea and Molly Holzschlag (2005) provide several creative examples of fixed positioning in their book, as well as on Shea's CSS Zen Garden Web site at http://www.csszengarden.com/.

RELATIVE POSITIONING

Relative positioning was originally intended to be used for scripting, as Hakon Lie and Bert Bos (1999), the designers of CSS, declared:

> The typical use of relative positioning is to position elements away from their normal position, without influencing the position of other elements. There aren't many reasons for doing that in a style sheet, and the main reason relative positioning exists at all is to provide a way for scripts to animate the text . . . . (216)

The code for a relatively positioned class looks like this:

```
.relative {position:relative; bottom:100px;
    left:100px;}
```

If I applied the relative class to one of my paragraphs, the result in a browser would look like Figure 11–6.

More recently, other uses for relative positioning have been developed beyond scripting. Both Eric Meyer (2003) and Dave Shea and Molly Holzschlag (2005) discuss ways of using relatively positioned elements as containers, or wrappers, for other elements. Remember that the absolutely positioned element is positioned according to its containing block, that is, the closest ancestor element that has been positioned. If there is no positioned ancestor element, the containing block is the page itself (i.e., html). With a relatively positioned containing block, you can absolutely position an element within another element rather than within the page itself, and you can create a centered layout with a fixed width much more easily than with absolute positioning alone.

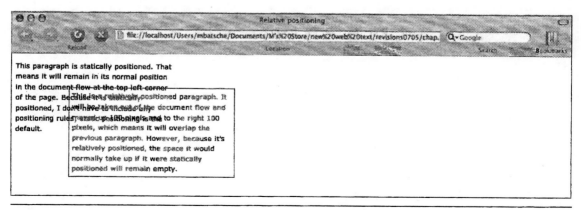

FIGURE 11–6   An example of relative positioning

Designers have begun using relatively positioned containing blocks that are centered on the page by setting the left and right margins at auto. These relatively positioned boxes can then contain other elements, both absolutely and relatively positioned. The resulting designs usually repeat background images placed within the body, whereas the content and navigation portions of the page have solid backgrounds that make them stand out against the background patterns. The site for the National Science Foundation (http://www.nsf.gov) uses relative positioning in this way (see Figure 11–7). The several parts of the Library of Congress American Memory collection (http://memory.loc.gov/ammem/)— a good source of public domain images—also use this type of design (see Figure 11–8).

A CSS for a typical layout using a relatively positioned containing block would look like this:

```
body {margin:0; padding:0;}
#container {position:relative; padding:10px;
    width:65%; margin:0 auto 0 auto;}
#linklist {position:absolute; top:0; left:0;
    width:20%; padding:10px;}
#content {position:relative; left:25%; width:70%;}
```

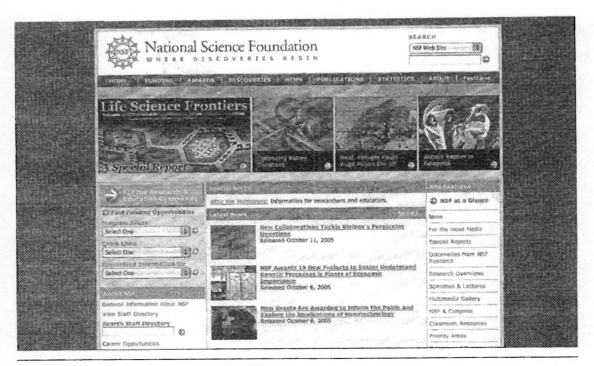

FIGURE 11–7   The National Science Foundation site

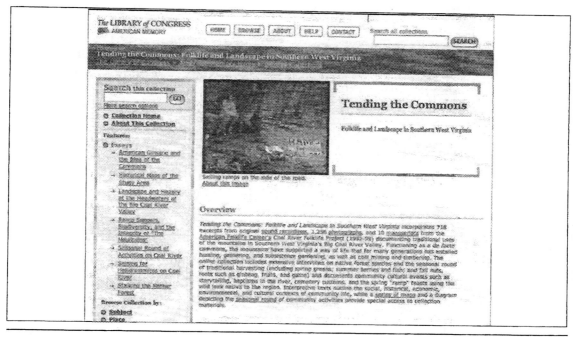

FIGURE 11–8 The Library of Congress American Memory collection

The container div would contain both the link list and the content div, thus opening first and closing last. The HTML would look like this:

```
<!DOCTYPE html PUBLIC "-//W3C//DTD XHTML 1.0
    Transitional//EN""http://www.w3.org/TR/xhtml1/DTD/
    xhtml1-transitional.dtd">
<html xmlns="http://www.w3.org/1999/xhtml">
<head>
    <link rel="stylesheet" href="relative2.css"
        type="text/css" />
    <title>Relative Positioning</title>
</head>
<body>
<div id="container">
<div id="content">
<p>
Both Eric Meyer (2003) and Dave Shea and Molly
Holzschlag (2005) discuss ways of using relatively
positioned elements as containers, or wrappers,
for other elements. Remember that the absolutely
positioned element is positioned according to its
containing block, that is, the closest ancestor
```

```
element that has been positioned. If there is no
positioned ancestor element, the containing block
is the page itself (i.e., html). With a relatively
positioned containing block, you can absolutely
position an element within another element rather
than within the page itself, and you can create a
centered layout with a fixed width much more easily
than with absolute positioning alone.
</p>
<p>
Designers have begun using relatively positioned
containing blocks that are centered on the page by
setting the left and right margins at auto. These
relatively positioned boxes can then contain other
elements, both positioned and unpositioned. The
resulting designs usually repeat background images
placed within the body, whereas the content and
navigation portions of the page have solid
backgrounds that make them stand out against
the background patterns.
</p>
</div>
<div id="linklist">
<p>
This menu will be placed inside the container div.
</p>
<ul style="list-style-type:none;">
    <li>Link</li>
    <li>Link</li>
    <li>Link</li>
</ul>
</div>
</div>
</body>
</html>
```

In a browser it would look like Figure 11–9.

Nested Boxes

You need not position the elements you place within your relative containers
unless you want a column layout. If you want a horizontal layout, you can place
unpositioned elements within containing elements both to position them and to
set their content off from the rest of the content. A simple nested box within a
larger box could have a style sheet like this:

```
body {margin:0px; padding:0px;}
.innerbox {padding:10px; margin:10px 0px 5px 0px;}
```

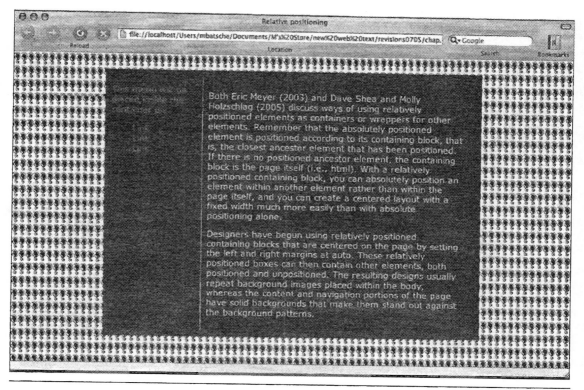

FIGURE 11–9   A relatively positioned container

```
#container {padding:10px; margin-top:5px; margin-
bottom:5px; margin-right:auto; margin-left:auto;
width:70%;}
```

The inner box here is set up as a class so that you can reuse it. The containing box has been centered, using the auto value. The nested box will be positioned by the normal flow of the content. Its width will be determined by the padding in the containing box, and its height will be determined by the height of the content. The HTML would look like this:

EXERCISE: RELATIVE POSITIONING

If any of your page designs would work best with a centered container, try setting them up with relative positioning. Remember to use a combination of both relative and absolute positioning to create your design; elements within the relatively positioned container must use either absolute or relative positioning.

```
<!DOCTYPE html PUBLIC "-//W3C//DTD XHTML 1.0
    transitional//EN""http://www.w3.org/TR/xhtml1/DTD/
    xhtml1-strict.dtd">
<html xmlns="http://www.w3.org/1999/xhtml">
<head>
        <title>Nested Boxes</title>
        <link rel="stylesheet" href="nested.css"
        type="text/css" />
</head>
<body>
    <div id="content">
        <p>You need not position the elements you
place within your relative containers unless you
want a column layout. If you want a horizontal
layout, you can place unpositioned elements
within containing elements both to position them
and to set their content off from the rest of the
content.</p>
            <div class="innerbox">
                <p>The inner box here is set up as a class
so that you can reuse it. The containing box has been
centered, using the auto value. The nested box will
be positioned by the normal flow of the content. Its
width will be determined by the padding in the
containing box, and its height will be determined by
the height of the content.</p>
            </div>
            <p>You need not position the elements you
place within your relative containers unless you
want a column layout. If you want a horizontal
layout, you can place unpositioned elements
within containing elements both to position them
and to set their content off from the rest of the
content.</p>
    </div>
</body>
</html>
```

In a browser it would look like Figure 11–10.

You can also use nested boxes to create simple navigation bars at the bottom or top of your page. The style sheet would look like this:

```
body {margin:0px; padding:0px;}
#content {padding:10px; margin-top:5px; margin-
    bottom:0px; margin-right:auto; margin-left:auto;
    width:70%;}
```

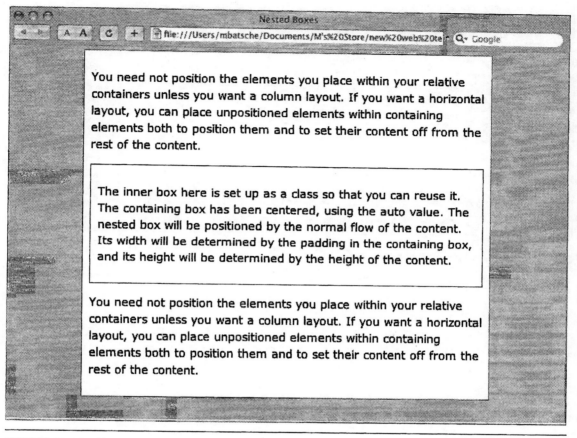

FIGURE 11–10 Nested boxes

```
#navbar {padding:0px 10px 0px 10px; margin-top:0px;
    margin-bottom:5px; margin-right:auto; margin-
    left:auto; width:70%;}
```

The HTML would look like this:

```
<!DOCTYPE html PUBLIC "-//W3C//DTD XHTML 1.0
    Transitional//EN""http://www.w3.org/TR/xhtml1/DTD/
    xhtml1-strict.dtd">
<html xmlns="http://www.w3.org/1999/xhtml">
<head>
    <title>Nested Navigation Bar</title>
    <link rel="stylesheet" href="nested2.css"
    type="text/css"/>
</head>
<body>
```

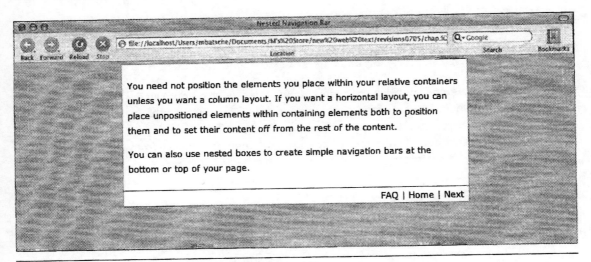

FIGURE 11–11   A Nested Navigation-Bar Box

```
<div id= "content">
    <p>You need not position the elements you place
within your relative containers unless you want
a column layout. If you want a horizontal layout,
you can place unpositioned elements within contain-
ing elements both to position them and to set their
content off from the rest of the content.</p>
<p>You can also use nested boxes to create simple
navigation bars at the bottom or top of your
page.</p>
    </div>
    <div id="navbar">
FAQ | Home | Next
    </div>
</body>
</html>
```

And in a browser it would look like Figure 11–11.

DISPLAY AND Z-INDEX

Display

Two other CSS properties can be applied to page layout: display and z-index. You'll see one of the uses of the display property with CSS navigation bars later in this chapter. **Display** is used to determine how an element will be rendered in a browser. It has a variety of possible values, some of which will be more useful to you than others. These are a few of the values for `display`:

❑ `inline`—If display is set to `inline`, the element will be displayed with an in-line box, that is, it won't begin on a separate line as a block-level element would.

❑ `block`—If display is set to `block`, the element will be displayed with a block-level box, that is, it will be placed on a new line.

❑ `none`—If display is set to `none`, the element will be invisible, as will any elements associated with that element as descendants. Thus, a paragraph in a div with a display set to `none` will also be invisible.

There are other values for display, particularly related to tables, that are beyond the scope of this discussion. If you're interested, you can find more information in CSS guides, such as those by Meyer (2003) or Musciano and Kennedy (2002).

Jeffrey Zeldman (2003) suggests using `display:block` with images to turn them into block-level elements. If you treat images as block-level elements, you won't have to use `<br/>` to place them on a separate line, and you'll have some of the advantages that used to come with placing images in their own table cells. Links can also be made of block-level images, as Zeldman demonstrates in his CSS buttons (discussed later in this chapter.) Eric Meyer (2003) points out that making links into block-level elements gives you the same effect that you'd have if you placed each link in its own div, thus allowing you to style the links as you would any other block-level element. They'll have automatic vertical spacing, and you can add margins or padding to move them in from the box edge.

However, in his *Programmer's Reference* Meyer (2001) cautions against going overboard with the display property "in a document language which already has a strong display hierarchy, such as HTML. Considerable havoc could result from setting all elements to block, for example; declaring everything to be inline could be just as bad" (123).

Z-Index

The **z-index** is a way to change the order of stacked block-level elements, using the z-axis. As you saw with background-image positioning, the x-axis runs left and right (i.e., horizontally), whereas the y-axis runs up and down (i.e., vertically). The z-axis is a third dimension, representing a line between the screen and the viewer's eyes. When you set the z-index for overlapping elements, you determine which one will be on top (i.e., closest to the viewer), and what the order of the others will be. The rule is straightforward: the larger the z-index number, the closer the element will be to the top of the stack.

Say you had an external style sheet with three ids for three divs:

```
#one {position:absolute; top:10px; left:10px;
    padding:10px; width:20%;}
```

```
#two {position:absolute; top:40px; left:40px;
    padding:10px; width:20%;}
#three {position:absolute; top:70px; left:70px;
    padding:10px; width:20%;}
```

Placed with HTML, the code would look like this:

```
<body>
<div id="one">
1
</div>
<div id="two">
2
</div>
<div id="three">
3
</div>
</body>
```

The boxes will be stacked in the order in which they occur since no z-index has been indicated. In a browser it would look like Figure 11–12.

Now let's add a z-index number to the boxes:

```
#one {position:absolute; top:10px; left:10px;
    padding:10px; width:20%; z-index:3;}
#two {position:absolute; top:40px; left: 40px;
    padding:10px; width:20%; z-index:2;}
#three {position:absolute; top:70px; left:70px;
    padding:10px; width:20%; z-index:1;}
```

With the same HTML code, the results would look like Figure 11–13.

You can stack elements using the z-index as long as they have the same parent element. The three divs in this example are all within the same parent element, the page itself. But if I had two divs, each with paragraphs inside, and I wanted to restack the paragraphs, I could change the stacking order only of the paragraphs that were in the same div.

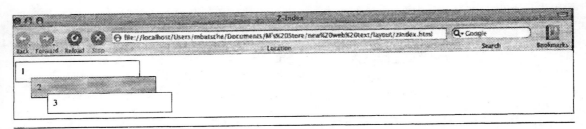

FIGURE 11–12   Stacking without a z-index

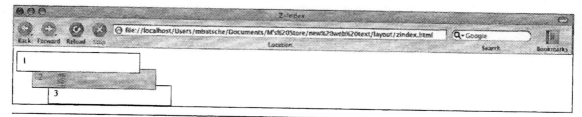

FIGURE 11–13   Stacking with a z-index

USING CSS LAYOUT FOR NAVIGATION

CSS layout provides a variety of possibilities for improved navigation using text rather than graphics. We'll discuss two of several possibilities: using CSS to create a vertical button bar and using CSS to create a horizontal tab bar.

Creating a Vertical Navigation Menu

CSS makes it possible to create a vertical menu bar that uses text buttons. As I pointed out in chapter 7, text is easier to revise than images, so a text button bar will simplify your maintenance even though the CSS layout will still look like image buttons. The following method is adapted from Jeffrey Zeldman's book (2003), *Designing with Web Standards*. It uses a combination of borders, background colors, and pseudoclasses to make a menu bar. For a more complex and detailed version, consult Zeldman's book.

The menu begins with an `id` called "navigation" and a `ul` that's defined as part of that `id`. The style rule uses `list-style:none` to remove the default bullets from the unordered list and sets the padding and border at zero for the list as a whole. It also establishes a top and bottom margin of fifteen pixels and a left and right margin of zero for the entire vertical button bar. Here are the styles that will create the bar itself:

```
#navigation ul{
list-style:none;
padding:0;
margin:15px 0;
border:0;}
```

Next comes the style rule for the list items that will be the buttons. This rule sets the text alignment and font, as well as the color and background color. The margin and padding are both set at zero, but each list item is given a black bottom border.

```
#navigation li{
text-align:center;
border-bottom:1px solid #000;
```

```
width:125px;
margin:0;
padding:0;
font:small/150% verdana, arial, helvetica, sans-serif;
color:#000;
background:#f93;}
```

Next comes the style rule for the list items that are links. Notice that the links are set to display:block. That property and value will mean that the links, which are in-line elements by default, will be displayed as block elements on separate lines. This rule also sets a left and right border for these list items and removes the default link underlining.

```
#navigation li a{
display:block;
font-weight:normal;
padding:0;
border-left:1px solid #000;
border-right:1px solid #000;
background: #f93;
color:#000;
text-decoration:none;}
```

Now comes the rule for the hover style:

```
#navigation li a:hover{
font-weight:normal;
background:#c90;
color:#fff;
text-decoration:none;}
```

A final style rule sets a top border for the top button, which won't have one otherwise:

```
#navigationtop{
border-top:1px solid #000;}
```

Here's the XHTML for the menu bar, linked to the external style sheet that includes all of these rules:

```
<!DOCTYPE html PUBLIC "-//W3C//DTD XHTML 1.0
    Transitional//EN""http://www3.org/TR/xhtml1/DTD/
    xhtml1-transitional.dtd">
<html xmlns="http://www.w3.org/1999/xhtml">
<head>
    <title>CSS Buttons</title>
    <style type="text/css">
@import url(button.css);
</style>
</head>
```

```
<body>
<div id="navigation">
<ul>
    <li id="navigationtop"><a href="http://
communication.utsa.edu/" title="UTSA Communication
Department">Communication Department</a></li>
    <li><a href="http://communication.utsa.edu/
mbatch/" title:"Margaret Batschelet's Web
Site">Batschelet Home Page</a></li>
    <li><a href="http://communication.utsa.edu/mbatch/
3413/" title="Writing for New Media">Communication
3413: Writing for New Media</a></li>
</ul>
</div>
</body>
</html>
```

In a browser it would look like Figure 11–14 (the middle button shows the hover effect).

Creating a Horizontal Navigation Menu

An unordered list can also be turned into a horizontal navigation bar, with effects similar to the tabs found on some image-based navigation bars. This method is adapted from Dan Cederholm's book (2004) *Web Standards Solutions*. For more details consult Cederholm's book.

The style rules begin with a rule for an id called "minitabs," which will be associated with the unordered list. This rule turns off the margins and padding, along with the default list indentation, and creates a bottom border.

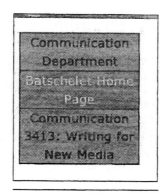

FIGURE 11–14   CSS
buttons

```
#minitabs {margin:0;
padding:0 0 20px 10px;
border-bottom:1px solid #696;}
```

The next rule uses the `display:inline` property and value to make the list items horizontal rather than vertical. It also removes the default bullets.

```
#minitabs li{
margin:0;
padding:0;
display:inline;
list-style:none;}
```

Another rule is added for the links themselves; it floats them to the left, as well as adding some padding and margins to the individual links.

```
#minitabs a{
float:left;
line-height:15px;
font-weight:bold;
margin:0 10px 4px 10px;
text-decoration:none;
color:#9c9;}
```

Now a rule is added for the active and hover states, which will add a thick underline when users pass a cursor across the link or click on it.

```
#minitabs a:hover, #minitabs a:active{
border-bottom:4px solid #696;
padding-bottom:2px;
color:#363;}
```

The HTML for the navbar is simple; the page is linked to a style sheet that includes all of these rules.

```
<!DOCTYPE html PUBLIC "-//W3C//DTD XHTML 1.0
    Transitional//EN""http://www.w3.org/TR/xhtml1/
    DTD/xhtml1-transitional.dtd">
<html xmlns="http://www.w3.org/1999/xhtml">
<head>
    <title>Horizontal Navigation Bar</title>
    <link rel="stylesheet" href="minitabs.css"
    type="text/css" />
</head>
```

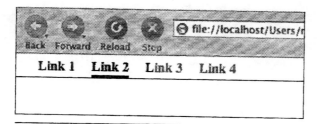

FIGURE 11–15 A horizontal link list

```
<body>
<ul id="minitabs">
    <li><a href="link1.html">Link 1</a></li>
    <li><a href="link2.html">Link 2</a></li>
    <li><a href="link3.html">Link 3</a></li>
    <li><a href="link4.html">Link 4</a></li>
</ul>
</body>
</html>
```

In a browser it would look like Figure 11–15 (Link 2 shows the hover style).

EXERCISE: CREATING NAVIGATION

If any of your page designs use either horizontal or vertical navigation bars, try using the templates described here. You can customize these designs by using your own colors, font, and so on.

BEYOND THE BASICS

The layouts I've shown you throughout this chapter are pretty rudimentary. In many cases I've used borders to show you where the boxes were; but in reality, you'd seldom have as many boxes on a single page as I've included—in the three-column layouts, for example. Nonetheless, these templates should give you some ideas about things to try. You can find more CSS layouts online in various CSS sites; try the CSS Zen Garden (http://www.csszengarden.com/), Dan Cederholm's SimpleBits site (http://www.simplebits.com/), Jeffrey Zeldman's site (http://www.zeldman.com/), and Eric Meyer's site (http://www.meyerweb.com/) to begin with. CSS layouts don't have to be boxy and uninspired—you can use Web standards and still come up with attractive, interesting layouts. And as a bonus, they'll be accessible to users with a variety of devices.

SIDEBAR: BROWSER SUPPORT FOR POSITIONING

There is more browser support for positioning now than there was a few years ago, but there are still some differences between browsers and some browser idiosyncrasies to be aware of. You can find the W3C's list of browsers supporting CSS at http://www.w3.org/Style/CSS/#browsers. In addition, the Position Is Everything site at http://www.positioniseverything.net/ keeps track of current developments in browser support and nonsupport. Finally, West Civ (http://www.westciv.com/style_master/ academy/browser_support/) provides tables that list the support of some browsers for various aspects of CSS.

PROCESS FOR DESIGNING PAGE TEMPLATES WITH CSS

Use the following steps to design your page templates using CSS positioning:

1. Look back over your rough drafts of the page templates for your site. Try to divide each template into its major parts.
2. After you've subdivided the templates, see whether you can set the design up as columns and rows or as centered horizontal segments.
3. Once you've identified the columns and rows, try matching them to one of the CSS templates in this or the previous chapter.

   ❏ If the design has only two columns or if the design requires a footer section, consider using floats.

   ❏ If the design has more than two columns or if the design risks falling apart if the window size is changed, consider using `position:absolute` for at least one of the columns.

   ❏ If the design includes elements that should remain on the screen at all times, consider using `position:fixed`.

   ❏ If the design would work well in a centered section with a patterned background, consider using `position:relative` for a containing block, with absolutely and relatively positioned segments inside it.

4. If your design includes either horizontal or vertical navigation menus, use or adapt any of the options suggested in this chapter.
5. Set up a style sheet for each template page, and link all the pages that will be using each template design to the appropriate style sheet.
6. Test each page in several Web browsers to make sure that your designs will show up the way you want them to.

CHAPTER SUMMARY

Positioning

❏ The `position` property determines whether elements are placed in the normal flow of the page or taken out of the flow and positioned in some other way. There are four possible values: `static`, `absolute`, `fixed`, and `relative`.

position: absolute

❏ Most positioned elements use absolute positioning. Absolutely positioned elements are removed from the document flow and positioned in a particular place on the page. The other elements on the page behave as if the absolutely positioned element isn't there.

❏ Absolutely positioned elements are positioned in their containing block, the nearest positioned ancestral unit. If there is no positioned ancestor, the element is positioned using the page itself as the ancestor.

❏ The ancestor of the absolutely positioned element is `html` rather than `body`. This means that margins and padding applied to the `body` selector will have no effect on an absolutely positioned element.

❏ Absolutely positioned elements must use two of four other properties to give coordinates: `top`, `right`, `bottom`, or `left`. The values stand for the distance away from the top, right, bottom, or left edge of the containing block within which the element is to be positioned.

❏ Multicolumn layout is the most common use of absolute positioning. Column layout uses a combination of margins, positioning, and document flow to line up columns.

❏ In a simple two-column layout, the menu box is positioned absolutely at one side of the page, whereas the content is given an extra-large margin so that it flows beside the menu.

❏ In a three-column layout, menu boxes are positioned absolutely on either side of the page, whereas the content box uses extra-wide side margins to flow between the two side boxes. The columns will have unequal lengths because length is determined by the content of the boxes.

❏ In a three-column layout with a banner across the top, you must adjust the top positioning of the two side boxes. The banner box will have a specified height and will be placed first within the HTML code.

position: fixed

❏ The containing block for an element with fixed positioning is the browser window; thus, a fixed element will always remain on-screen. However, other elements ignore fixed elements, so the spacing of other elements must be adjusted to avoid overlap.

`position: relative`

❑ Relative positioning was originally designed for scripting purposes, but it can also be used to position an absolute element within something other than the page itself.

❑ Relatively positioned elements can be used as containers for other elements in order to center a design on the page.

Display and Z-Index

❑ The `display` property determines how an element will show up in a browser.

`display:block` can be used to turn images into block-level elements and place links in their own boxes so that they can be used to create a graphic-style menu.

`display:inline` can be used to convert block-level elements to in-line elements.

`display:none` makes elements invisible.

❑ The `z-index` property allows you to change the stacking order of elements.

❑ The `display` property can be combined with other CSS rules to create both vertical and horizontal navigation bars, which can be maintained more easily than graphic links and can change color when visited.

CSS PROPERTIES IN THIS CHAPTER

PROPERTY	VALUES	EFFECT
height	Length value, percentage, auto	Sets height of content within block-level element
position	absolute	Removes element from document flow and positions it in relation to a containing block (other elements behave as if absolutely positioned element isn't there)
position	fixed	Like absolute positioning, uses window as containing block; keeps element on-screen while content flows behind it
position	relative	Removes element from document flow but preserves space it originally filled
top	Length value, percentage, auto	With `position` property, sets the inset distance from the top edge of the containing block for the positioned element
right	Length value, percentage, auto	With `position` property, sets the inset distance from the right edge of the containing block for the positioned element
bottom	Length value, percentage, auto	With `position` property, sets the inset distance from the bottom edge of the containing block for the positioned element

Continued

Continued

PROPERTY	VALUES	EFFECT
left	Length value, percentage, auto	With `position` property, sets the inset distance from the left edge of the containing block for the positioned element
display	inline, block, none	Determines how element will be rendered in the browser
z-index	number	Determines the stacking order of elements within the same containing block

SOURCES

Briggs, O. 2002. Boxes, boxes, boxes. Chap. 8 in *Cascading Style Sheets: Separating content from presentation.* Birmingham, UK: Glasshaus.

Cederholm, D. 2004. *Web standards solutions: The markup and style handbook.* Berkeley, CA: Friends of Ed/Apress.

Lie, H., and B. Bos. 1999. *Cascading Style Sheets: Designing for the Web,* 2nd ed. New York: Addison Wesley.

Meyer, E. 2001. *Cascading Style Sheets 2.0: Programmer's reference.* New York: Osborne.

———. 2003. *Eric Meyer on CSS.* Indianapolis, IN: New Riders.

Musciano, C., and B. Kennedy. 2002. *HTML and XHTML: The definitive guide,* 5th ed. Sebastopal, CA: O'Reilly.

Shea, D., and M. Holzschlag. 2005. *The Zen of CSS design: Visual enlightenment for the Web.* Berkeley, CA: Peachpit Press.

Zeldman, J. 2003. *Designing with Web standards.* Indianapolis, IN: New Riders.

Using Multimedia

MULTIMEDIA

The W3C regards graphic and text files as multimedia elements; however, the term *multimedia* is more frequently applied to elements such as audio and video files, as well as animation. All of these formats can be added to Web sites.

One of the great advantages of the Web is this availability of information in a variety of formats. But even though these formats have obvious advantages (e.g., hearing a speech or a piece of music is obviously superior to only reading about it), there are also several problems involved with using multimedia.

In this chapter we'll discuss methods of inserting other media types into Web pages, including the currently chaotic state of code available for embedding media files. We'll also look at the variety of file types available for both sound and video and the pros and cons of each. Finally, we'll look at Flash animation, a format that can incorporate both sound and video, as well as text and graphics. First, however, we need to consider some general principles about using multimedia.

Using Multimedia

Multimedia files can greatly enhance your delivery of information and can also entertain your users and attract their attention. But they do so at a cost. Many sound and video files are quite large and thus require extra time to download. Users without broadband connections may be particularly disadvantaged by the wait required for multimedia files. Streaming media technology can reduce that wait somewhat, but streaming files require browsers with media players that can deliver media in that format as well as hardware and software to create the files themselves.

Methods of creating sound and video files are beyond the scope of this chapter. Although it has become easier to create both digital audio and video than it was even a few years ago, you would still need recording equipment

SIDEBAR: MULTIMEDIA TERMS

Embedded files: Files that actually play on the Web page itself rather than in a separate application.

MIME: Multipurpose Internet mail extension. A protocol developed to allow various file types to be sent as attachments to e-mail messages. MIME is now used by Web browsers to identify multimedia file types.

Param: A tag that supplies parameters that provide additional information about media files, including type and playback options.

Plug-in: A helper application that enables a Web browser to display or execute a multimedia file. Both Windows Media Player and Quicktime are plug-ins that allow Web browsers to play a variety of media files.

and software for editing and playback. Instead, you can buy audio and video files to use on your pages, but you'll be limited to the file types used to create them.

The general principal here is simple: don't use audio or video unless you have a good reason for doing so. If you've ever sat through a long download for a cheesy audio track you didn't want to hear or a video that didn't turn out to be worth waiting for, you can appreciate the need to limit your multimedia files to essentials. And don't just dump an audio or video track on your page without giving your users a choice about listening to or watching it. Internet Explorer has a `bgsound` tag that allows a sound file to begin playing as soon as a page is loaded. We won't cover the tag since it's browser specific, but consider the potential annoyance if a sound file begins playing unexpectedly in a computer lab or an office where others are trying to work. The user who inadvertently accessed the page with that file isn't at fault; the designer is.

Plug-ins and Media Players

Netscape Navigator was the first Web browser to provide support for multimedia files, and it did so by using helper application software that existed outside the browser itself, that is, **plug-ins.** The basic system developed by Netscape is still in use with modern Web browsers, although it has undergone several modifications since the early days of the Web.

When a browser encounters something other than text or a graphic file on a Web site, it identifies the file format by its extension and its MIME type. MIME, which stands for multipurpose Internet mail extension, is a protocol that was established to allow different file types to be sent as attachments to e-mail. However, MIME has now become a standard format for exchanging files across the Internet. The **MIME type** is specified in a two-part code, using a forwardslash. In fact, you've already used a MIME type with CSS,—`type="text/css"`. The first part of the code, `text`, specifies the MIME type; the second part, `css`, is the subtype. We'll cover a method of assigning MIME types to your multimedia files as attributes later in this chapter.

By now, most browsers can identify a wide range of files by their extensions and MIME type. In order to execute the files, however, the browser may need to install a plug-in, a helper application developed by a software company to allow multimedia files to be displayed or executed within the browser. For example, Macromedia has developed a Flash Player, which allows Flash movies to be played within a browser. If the browser doesn't have the Flash Player installed or if the version of the Flash Player is incorrect, the user will need to download the current Flash Player in order to see Flash files. Most modern browsers include a wide variety of preinstalled plug-ins to avoid the delay of locating and downloading the necessary helper applications, but new file formats are constantly being developed, especially for multimedia files. Your users may have to download new plug-ins to accommodate any cutting-edge technology you decide to use.

Multimedia and Copyright

The issue of copyright occurs with multimedia files just as it did with graphics. The long-running dispute between the Recording Industry Association of America (RIAA) and the Napster file-sharing utility has highlighted some of the issues involved. In essence, you cannot simply post a digital version of your favorite song or embed it on your page as background music unless you pay for the right to use that song. Songs and films involve many more rights than photographs and drawings. For example, the composer of a song has one set of rights, whereas the recording artist and the recording company have another. With films and videos, not only does the owner of the film have rights, but the actors involved may have separate rights, and clearance must be obtained for all of them. Thus, although you might be able to download a digital version of a movie clip from another site, you can't place that clip on your own site without running the risk of copyright infringement.

Occasionally, artists make their songs or videos available over the Web for promotional purposes, and collections of royalty-free music and videos are available for purchase, like clip art collections. If you have access to audio and video hardware and software, you can, of course, create your own audio and video. But you cannot use copyrighted audio or video material on your Web site without obtaining permission and possibly paying royalty fees.

SOUND FORMATS

In chapter 6 you saw that only three graphic formats (GIF, JPEG, and PNG) were currently used within Web browsers. The situation is much more complex in terms of audio and video files; several formats are available for each. And as with graphic files, there are advantages and disadvantages to each format that you should be aware of.

Sound files are available in a variety of formats, but some of those formats are platform-specific, or playable only on certain media players, which limits their applicability on Web pages. The following descriptions can give you some idea of the pluses and minuses of the various audio formats available.

Basic Audio

Basic audio (`.au`) from Sun Microsystems has universal support among browsers and platforms. However, it supports only 8-bit sound, which may not be enough to reproduce music files effectively (MP3 files use 128-bit sound). The MIME type for basic audio is `audio/basic`.

Audio Interchange File Format

Audio interchange file format (`.aif`) was originally developed by Apple for Macintosh computers. Although AIFF files are now supported by other platforms and are supported by some browsers, they are less well supported than other audio file types. In addition, AIFF files tend to be larger than some other audio files because they cannot be compressed. The MIME type for AIFF files is `audio/x-aiff`.

Moving Picture Experts Group Layer-3 Audio

This format, known as MP3 (`.mp3`), was originally developed for video, but the files had both high quality and excellent compression, making them extremely popular for digital music file sharing. Currently, many programs exist for "ripping" files from audio CDs and converting them to MP3 files, which can be easily loaded onto Web pages. However, remember that copyrighted music files cannot be used on your Web site without permission. MP3 is one of the most popular and best-supported audio formats on the Web. The MIME type for MP3 files is `audio/mpeg`.

Musical Instrument Digital Interface

The musical instrument digital interface (`.mid`) has been in use since the early eighties, when it was developed by the music industry to be used with electronic synthesizers and to transfer music files to computers. The format remains popular for composing and editing music; MIDI files are fairly small and universally supported. The MIME type for MIDI files is `audio/mid`.

RealAudio

RealAudio (`.ra`), a proprietary format created by RealNetworks, is a primary format for streaming audio. RealAudio files can be played only by RealPlayers, unlike other audio files, which can be played by QuickTime, Windows Media Player, and other media software. However, RealPlayers are widely used and are installed in many Web browsers. The MIME type for RealAudio files is `audio/x-pn-realaudio`.

Waveform Audio File Format

The Waveform Audio File Format (.wav) was developed by Microsoft and IBM to be a standard audio format for PCs. WAV files are widely used across platforms and browsers, although their quality suffers with compression. Thus, WAV files are used less frequently for music than MP3 files. However, WAV files remain popular for sound effects and other small sound files. The MIME type for WAV files is `audio/x-wav`.

Windows Media Audio

Windows Media Audio (.wma) is a proprietary format developed by Microsoft for the Windows Media Player. The files are also available in a streaming format with the .asf extension (i.e., active streaming file). These files can be played only by the Windows Media Player, but the player is distributed with the Windows OS and is also available for Macintosh. Many recent browser versions have the Windows Media Player installed. The MIME type for .wma files is `audio/x-ms-wma`; for `asf` files, it's `video/x-ms-asf`.

Selecting a Sound Format

Unless you create your own sound files or have access to a program that can reformat existing sound files, you may be limited to the format of the file you need to use. However, if you have a choice, the best format for audio depends on the type of audio file to be used. You'll probably want to avoid proprietary formats, if possible, in order to reach the widest number of users. Among the nonproprietary formats, MP3s have become the standard for digital music because of the quality of their reproduction and their small file size. WAV files are still widespread and are good for sound effects, and MIDI files are still used for electronic music and original compositions.

VIDEO FORMATS

No matter which format is used, video files are usually quite large, larger than sound files. Streaming video can help, but even so, you should limit your videos to short clips unless your users have high-speed connections. Flash (.swf) is common video format that I'll discuss later in this chapter.

Audio Video Interleave

Audio Video Interleave (.avi) was developed by Microsoft. Although it is widely used on the Web, support by other platforms and browsers is not consistent. Moreover, Microsoft is replacing AVI with another format, Windows Media Video (WMV). Still, many existing video clips use AVI. The MIME type for AVI files is `video/x-msvideo`.

Moving Picture Expert Group

The Moving Picture Expert Group (.mpg) developed the MP3 file format described earlier. Their video format offers similar advantages: it has high quality and a small file size. MPEGs are supported by most browsers and platforms. The MIME type for MPEG files is video/mpeg.

QuickTime

The QuickTime (.mov) format was developed by Apple and has now gone through several generations. Although originally a Macintosh-only program, QuickTime is now supported by most platforms and browsers. It has good compression and quality. The MIME type for QuickTime is video/quicktime.

RealVideo

Like RealAudio, RealVideo (.ram) is a proprietary format owned by RealNetworks. Because RealVideo files are designed to be streamed, they sometimes have reduced quality. As with RealAudio, these files can be played only by the RealPlayer. The MIME type for RealVideo is video/vnd.rn-realvideo.

Windows Media Video

Windows Media Video (.wmv) is a proprietary format from Microsoft; it can be played only on the Windows Media Player. The downloadable version of these files is saved with the .wmv extension, whereas the streaming version is saved with the .asf extension. The MIME type for .wmv files is video/x-ms-wmv; for .asf files it's video/x-ms-asf.

Selecting a Video Format

As with audio files, unless you create your own video files, you'll probably be limited to the format of the file you need to use. However, if you have a choice, video formats should be judged on the basis of their compression capability, even more than sound files. Given the size of most video files, you'll need to make sure that you're using the smallest version that will serve your needs; otherwise, the download time may simply be more than most users will tolerate. MPEG and QuickTime both have good compression formats, and both are widely available. AVI files may work for short video elements.

MEDIA PLAYERS

The most popular plug-ins provide access to stand-alone **media players**. The most popular players are usually available within your browser or for download to your computer. These are some of the most widely used media players.

QuickTime

QuickTime was originally created as a plug-in for Macintosh systems but is also available for Windows platforms. Its popularity is based on its support for over 200 file types and its cross-platform capabilities.

RealPlayer

RealPlayer is produced by RealNetworks and is particularly important for streaming media. The player supports a number of multimedia formats but is also the only player to support the proprietary RealAudio and RealVideo formats.

Windows Media Player

Windows Media Player is part of the Windows operating system and is thus widely available. A new version is also available for Macintosh machines. Windows Media Player supports most popular audio and video formats.

Flash Player

The Flash Player is used exclusively for Flash files and is loaded into most popular browsers. Your browser must have the Flash Player plug-in to render Flash files.

PLACING MULTIMEDIA ON YOUR PAGE

In this discussion we'll concentrate on QuickTime and Windows Media Player files, the two most common file formats for audio and video files. The methods for placing audio and video files on your pages are the same, regardless of the nature of the file. You can link to the files, or you can embed them. If you link to a file, you provide your user with a choice about whether to download the file, but the file itself will be less integrated into your page, and you'll have no control over the technology used to play the file. If you embed a file, you can have more control over integrating it into your page and specifying players to be used, but your user will have less control over the download and playback.

Linking to a Multimedia File

You use the same basic tags to link to an audio or video file that you use to link to a page—the anchor tags. If you wanted to provide a link to a particular audio file, for example, the code might look like this:

```
<a href="audiofile.au"> Founder's Day Speech</a>
```

Any styles that you applied to other links would also be applied to the links to multimedia files. Thus, the links to your multimedia files would look exactly like the links to other files on your site. Your users can decide whether to click on the multimedia link and download the file. If they decide to do so, the file will be played by whatever plug-in they have installed on their systems or within their browsers, provided the plug-in recognizes the file type. You can include information about the file size and type within the link, which will give your users some idea of what kind of download they'll be getting. Here's an example:

```
<a href="audiofile.au"> Founder's Day Speech
    (4 megabytes; basic audio (au) file) </a>
```

You can also indicate your preferred media player, although some users may resent being directed away from their media player of choice. It may be best to limit such directions to cases in which a particular file type can be played only by a particular media player.

Most browsers will open a linked media file in a new window with the browser's embedded player. If the correct plug-in is not available in the browser, the linked file will be opened in a player on the user's computer if one is installed. If neither the browser nor the computer has the correct player available, most browsers will allow the user to download the file or choose another program in which to open it. However, no matter which plug-in is or isn't available, the browser will not download the file unless the user chooses to do so.

Embedding a Media File

Instead of linking to an external file, you can choose to embed your file on the page itself. Embedded elements are actually played on the Web page itself rather than being downloaded and played on a separate page in the browser. The code for embedding files is currently in what can charitably be called a state of transition. Although the W3C specifies a relatively clear-cut method of embedding files, it has yet to be correctly implemented in all browsers.

A little history may help to explain the problems. When Netscape developed the plug-in system, it also developed the embed tag. Because Netscape dominated the browser market, other browsers supported the embed tag as well. However, the W3C created its own object tag, which can embed more types of files than the embed tag can and is thus considerably more powerful. In order to compete with Netscape, Internet Explorer began supporting the object tag, but not in the way specified by the W3C. Then and now, Explorer supports the object tag only when it is used with Microsoft's proprietary ActiveX controls, which are not available to other software producers. In addition, beginning with Internet Explorer 5.5 for Windows, Explorer no longer supported the embed tag. Netscape 6 fully supported the object tag following the W3C specification, but since Netscape cannot read ActiveX controls, it does not support the object tag in the way it's used by Internet Explorer.

The W3C specification for the object element could provide a solution for this dilemma. As Elizabeth Castro (2003) points out, "The object element is designed to be nested in order to allow several options to browsers. If the first object is unreadable, they can ignore it and try the next until they find one they understand" (295). This should make the `object` tag the ideal solution to the problem. However, Internet Explorer does not follow the W3C specification for nested `object` tags; instead of ignoring the `object` elements it can't understand, Explorer places large, empty boxes on the page for the unreadable elements.

Given this lack of compatibility, there are two options for dealing with this problem. On the one hand, you can argue that XHTML supports only the `object` element and that other options, such as the `embed` tag, have been deprecated. Thus, you can use the `object` tag exclusively and hope that the next version of Internet Explorer will follow the W3C specifications more exactly. As Don Gosselin (2004) points out, "Keep in mind that Web pages you write with the `<bgsound>` and `<embed>` elements will not be well formed, even if you use the Transitional DTD" (549).

The alternative is to use a combination of standards-compliant code and deprecated tags. Netscape and other browsers still support the `embed` tag for backward compliance; Explorer will ignore the `embed` element if it's provided with the `object` element using ActiveX elements. Using both `object` and `embed` is hardly an ideal solution since it goes against the ideal of standards-compliant code, but it is the only way to work around the current situation unless you're willing to put up with the empty boxes supplied by Explorer for the `object` elements it doesn't understand.

EMBEDDING A FILE WITH OBJECT

The exact code you'll use with the `object` tag will vary, depending on the kind of media file and media player you're working with. Let's begin with a relatively simple example. As you saw in chapter 6, the `object` tag can be used to place images as well as other media. The format looks like this:

```
<object data="compwriter.gif" type="image/gif"
    width="50%" height="50%">
</object>
```

As I said earlier, the `type` attribute refers to the MIME type of the file. The `data` attribute gives the file name. Height and width, however, mean something different in the `object` tag as compared to the `img` tag. If `height` and `width` are given as percentages, as they are here, the values refer to how much of the browser window the object should take up—in this case 50 percent of the height and the width. In a browser it would look like Figure 12–1.

To maintain the graphic's correct proportions, you'd use the pixel dimensions, as you did with the `img` tag, like this:

```
<object data="compwriter.gif" type="image/gif"
    width="133" height="144">
</object>
```

The graphic would then look like Figure 12–2.

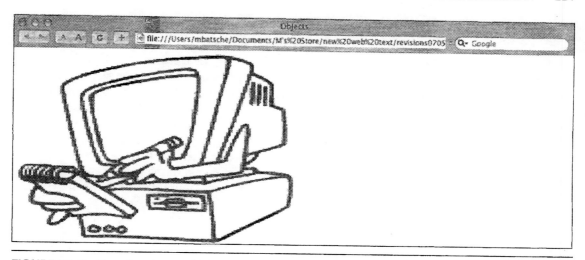

FIGURE 12–1   Use of `object` with 50 percent height and width

You should also supply text between `<object>` and `</object>` that can be displayed if the browser can't display the file itself. The code would look like this:

```
<object data="compwriter.gif" type="image/gif"
    width="133" height="144">
Your browser cannot display this object; this graphic
    is a picture of a writing computer.
</object>
```

For more complex files, the `object` tag may require other attributes beyond the `data`, `type`, `height`, and `width` attributes described here, including the following:

❑ `classid`—The `classid` provides an identifying number for the object, such as a plug-in, an ActiveX control, or a Java applet. For example, the

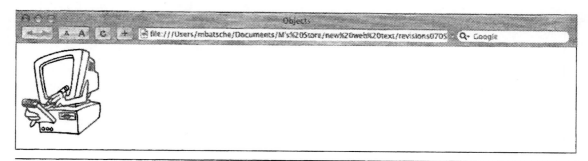

FIGURE 12–2   Use of `object` with pixel height and width

ActiveX `classid` for Explorer's control for QuickTime movies is `"clsid:02BF25D5-8C17-4B23-BC80-D3488ABDDC6B"`. As you might guess, the `classid` must be typed exactly like that, complete with hyphens and without spaces. (To locate the `classid` for a given plug-in, you can contact the producer of the plug-in itself.) The `classid` is required for Internet Explorer for Windows because it supplies the Acitve X control. However, other browsers will ignore an `object` tag that contains a `classid` or `codebase` attribute. For those browsers you'll have to supply the embed work-around described later.

❑ `codebase`—The `codebase` is a URL that provides the location for the plug-in you're referencing in your `classid`. If the plug-in is not installed in the user's browser, the `codebase` reference will send the browser to the right location to find it. Although not required, the `codebase` attribute is a good one to include with the `classid`.

❑ `name`—Using a `name` attribute will apply a unique name to an object, which can be helpful in scripting.

Using Param

More complex files using the `object` tag may also require the `param` tag. *Param* stands for parameter, and the `param` element identifies additional information required to run a particular file. The parameters required for a file vary, depending on the file and player type. For example, you may need to supply the file name and information about the controls to be displayed, as well as whether the file is to begin playing immediately and to repeat, or loop. `param` is an empty tag like `<br />` and `<link />`, which means in XHTML it's written `<param />`.

`param` can take several attributes: `type` sets the MIME type, and `id` creates a unique id. `value` and `valuetype` set the parameter's value and the value's type—`data`, `ref` for a URL, and `object`, which refers to another object on the page. Although `value` is usually required, `valuetype` may not be.

`name` indicates the kind of parameter involved. Several values are possible, depending on the kind of file and player the object refers to. For example, the Windows Media Player uses `<param name="autostart">`, which causes the file to begin or not begin playing, depending on whether you include `value="true"` for yes or `value="false"` for no. For QuickTime the equivalent name is `<param name="autoplay">`.

Table 12–1 includes some common parameters for Windows Media Player; Table 12–2 lists common parameters for QuickTime.

Embedding a Sound File

Now let's see how the entire process works to embed a sound file. First, be sure to put your `object` tag at the end of the page file so that the user won't have to wait while a large media file downloads before seeing the rest of the page. If you want the sound file to play automatically when the page loads and you're using Internet Explorer, the code for Windows Media Player would be the following:

TABLE 12–1   Windows Media Player Parameters

PARAMETER	DESCRIPTION	FORMAT
filename	Specifies the name of the file	`<param name="filename" value="mysound.au" />`
autostart	Determines whether file starts playing automatically when loaded	`<param name="autostart" value="true" />` or `<param name="autostart" value="false" />`
showcontrols	Specifies whether Media Player controls will show	`<param name="showcontrols" value="true" />` or `<param name="showcontrols" value="false" />`
clicktoplay	Determines whether file plays when user clicks in display	`<param name="clicktoplay" value="true" />` or `<param name="clicktoplay" value="false" />`
animationstart	Determines whether Microsoft animation show executes while file is loaded	`<param name="animationstart" value="true" />` or `<param name="animationstart" value="false" />`
transparentatstart	Determines whether object is transparent while loading	`<param name="transparentatstart" value="true" >` or `<param name="transparentatstart" value="false" >`

```
<object id="MediaPlayer1"
    classid="CLSID:22D6F312-B0F6-11D0-94AB-
    0080C74C7E95"
codebase="http://activex.microsoft.com/activex/
    controls/mplayer/en/nsmp2inf.cab#Version=6,4,5,715"
    height="20" width="150">
<param name="filename" value="sound.mp3" />
<param name="autostart" value="true" /> Your Web
    browser can't load this object. This object is a
    sound file playing the theme song for our Web
    site.</object>
```

To do the same thing in QuickTime (in Internet Explorer), the code would be this:

```
<object classid="clsid:02bf25d5-8c17-4b23-bc80-
    d3488abddc6b"
codebase="http://www.apple.com/qtactivex/qtplugin.cab"
    width="150" height="20">
<param name="src" value="sound.mp3" />
```

TABLE 12–2   QuickTime Parameters

PARAMETER	DESCRIPTION	FORMAT
src	Specifies the name of the file	`<param name="src" value="mysound.au" />`
autoplay	Determines whether file starts playing automatically when loaded	`<param name="autoplay" value="true" />` or `<param name="autoplay" value="false" />`
controller	Specifies whether QuickTime controls show	`<param name="controller" value="true" />` or `<param name="controller" value="false" />`
loop	Allows movie to repeat or to play alternately forward and backward (palindrome)	`<param name="loop" value="true" />` or `<param name="loop" value="false" />` or `<param name="loop" value="palindrome" />`
scale	Scales movie to fit in a rectangle or adjust aspect ratio or set scaling factor	`<param name="scale" value="ToFit" />` or `<param name="scale" value="Aspect" />` or `<param name="scale" value="n" />` (substitute for n the number by which the original height and width of the movie are to be multiplied)

```
<param name="autoplay" value="true" /> Your Web
   browser can't load this object. This object is a
   sound file playing the theme song for our Web
   site.</object>
```

SIDEBAR: WIDTH AND HEIGHT WITH VIDEO AND AUDIO FILES

You'll want to add height and width measurements to your video file tags in order to set the size of the box in which the video will be presented. As you saw previously with the GIF image, using percentages sets the display box to a percent of the browser window. You can also set the box to the dimensions of the video itself; the dimensions should be available in the program used to create the video or in your media player. You can set the box to the exact size of the video or add a few pixels to make the box slightly bigger or accommodate a frame.

It may seem odd to add width and height measurements to audio, something that has no width and height itself. However, there are good reasons for adding minimal width and height measurements: you'll need to provide enough space to show the controller for playing the audio file. The rule of thumb from the QuickTime developers is to provide a minimum of 150 pixels in width and a minimum of 16 pixels in height to accommodate the audio controller.

If you didn't want the sound file to play as soon as the page loaded, the value for `autostart` and `autoplay` would be `false`, and you'd probably want to add controls for the sound using `showcontrols` or `controller`.

Both of these files would play in Internet Explorer for Windows, but **they wouldn't play in any other browser** because no other browser could read the ActiveX controls stored in the `classid` and `codebase`. The standards-compliant way around this is to nest another `object` tag within the `object` tag with the ActiveX code. It would look like this for the Windows Media Player sound:

```
<object id="MediaPlayer1"
    classid="CLSID:22D6F312-B0F6-11D0-94AB-
    0080C74C7E95"
codebase="http://activex.microsoft.com/activex/
    controls/mplayer/en/nsmp2inf.cab#Version=6,4,5,715"
    height="20" width="150">
<param name="filename" value="sound.mp3" />
<param name="autostart" value="true" />
<object data="sound.mp3" type="audio/mpeg"
    width="150" height="20">
<param name="filename" value="sound.mp3" />
<param name="autostart" value="true" />
Your Web browser can't load this object. This object
    is a sound file playing the theme song for our
    Web site.
</object>
</object>
```

And it would look like this for the QuickTime sound:

```
<object classid="clsid:02BF25D5-8C17-4B23-BC80-
    D3488ABDDC6B" width="150"height="20"
codebase="http://www.apple.com/qtactivex/
    qtplugin.cab">
<param name="src" value="sound.mp3" />
<param name="autoplay" value="true" />
<object data="sound.mp3" type="audio/mpeg"
    width="150" height="20">
<param name="src" value="sound.mp3" />
<param name="autoplay" value="true" />
Your Web browser can't load this object. This object
    is a sound file playing the theme song for our
    Web site.
</object>
</object>
```

Embedding a Video File

You embed a video file in more or less the same way you embed an audio file, but with these examples we'll use controls rather than having the file start as soon as it's loaded. For a Windows Media Player video, the code would look like this:

```
<object id="MediaPlayer1" width="180" height="140"
    classid="CLSID:22D6F312-B0F6-11D0-94AB-
    0080C74C7E95"
codebase="http://activex.microsoft.com/activex/
    controls/mplayer/en/nsmp2inf.cab#Version=6,4,5,715"
        type="application/x-oleobject">
    <param name="filename" value="film.avi" />
<param name="showcontrols" value="true" />
<object data="film.avi" type="video/avi"
    width="180"height="140">
<param name="filename" value="film.avi" />
<param name="autostart" value="false" />
Your Web browser can't load this object. This object
    is a media player file showing the Founder's Day
    address.
</object>
</object>
```

For a QuickTime video, the code would look like this:

```
<object classid="clsid:02BF25D5-8C17-4B23-BC80-
    D3488ABDDC6B" width="180"height="120"
codebase="http://www.apple.com/qtactivex/
    qtplugin.cab">
<param name="src" value="film.mov" />
<param name="controller" value="true" />
<object data="film.mov" type="video/quicktime"
    width="180"height="120">
<param name="src" value="film.mov" />
<param name="autoplay" value="false" />
Your Web browser can't load this object. This object
    is a QuickTime file showing the Founder's Day
    address.
</object>
</object>
```

These nested `object` tags would play effectively in most browsers (including Internet Explorer for Macintosh), but in Internet Explorer for Windows, you'd end up with extra boxes for the embedded `object` tags, which probably wouldn't look the way you intended your page to look. Given the current state of implementation for the `object` element, you might want to use the work-around method described next until browser support for the `object` specifications is more widespread.

EMBEDDING A FILE WITH OBJECT AND EMBED

Combining the `object` and `embed` tags is not standards-compliant code because it uses a deprecated tag (`embed`). However, it is currently the only way to compensate for the different ways in which `object` has been implemented in Internet Explorer for Windows and other browsers. In fact, according to the QuickTime developer's guide, embedding the `embed` tag within the `object` tag is the recommended syntax for using QuickTime in any browser.

This is the basic syntax for a QuickTime movie:

```
<object height="140" width="180" classid="clsid:
    02bf25d5-8c17-4b23-bc80-d3488abddc6b"
codebase="http://www.apple.com/qtactivex/
    qtplugin.cab">
<param name="src" value="film.mov" />
<param name="autoplay" value="true" />
<embed height="140" width="180" type="video/quick-
    time" pluginspace="http://www.apple.com/
    quicktime/download/"
src="film.mov"
autoplay="true">
</embed>
</object>
```

`embed` takes slightly different syntax from `object`; rather than the `param` tag, `embed` uses the name of the parameter itself. Thus, in this example `object` uses `<param name="src" value="film.mov" />`, whereas `embed` uses `src="film.mov"` as an attribute within the `embed` tag. If you add any more parameters to the `object` tag, you'll also add them as attributes to the `embed` tag so that each part of the code contains the same variables.

Here's another example, this time for Windows Media Player:

```
<object id="MediaPlayer1" width="180" height="140"
    classid="CLSID:22D6F312-B0F6-11D0-94AB-
    0080C74C7E95"
codebase="http://activex.microsoft.com/activex/
    controls/mplayer/en/nsmp2inf.cab#Version=6,4,5,715"
        type="application/x-oleobject">
    <param name="filename" value="film.avi" />
<param name="autostart" value="false" />
<param name="showcontrols" value="true" />
<embed height="140" width="180" type="application/
    x-mplayer 2" pluginspace="http://www.microsoft.
    com/Windows/MediaPlayer/"
src="film.mov" autostart="false" showcontrols="true">
</embed>
</object>
```

Notice that I've added another parameter, "showcontrols," which appears in a parameter tag in the object tag and as an attribute in the embed tag.

The same syntax is used with sound files; here's an example for QuickTime:

```
<object height="20" width="150" classid="clsid:
    02bf25d5-8c17-4b23-bc80-d3488abddc6b"
codebase="http://www.apple.com/qtactivex/
    qtplugin.cab">
<param name="src" value="sounds.mp3" />
<param name="autoplay" value="true" />
<embed height="20" width="150" type="audio/mpeg"
    pluginspace="http://www.apple.com/quicktime/
    download/"
src="sounds.mp3"
autoplay="true">
</embed>
</object>
```

And here's an example for Windows Media Player:

```
<object id="MediaPlayer1" width="150" height="20"
    classid="CLSID:22D6F312-B0F6-11D0-94AB-
    0080C74C7E95"
codebase="http://activex.microsoft.com/activex/
    controls/mplayer/en/nsmp2inf.cab#Version=6,4,5,715"
        type="application/x-oleobject">
    <param name="filename" value="sounds.mp3" />
<param name="autostart" value="false" />
<param name="showcontrols" value="true" />
<embed height="150" width="20" type="audio/mpeg"
    pluginspace="http://www.microsoft.com/Windows/
    MediaPlayer/"
src="sound.mp3" autostart="false" showcontrols="true">
</embed>
</object>
```

As new versions of browsers become available, this method will become obsolete, and you will be able to use the object tag as it was intended to be used. However, for the moment this method is the most reliable way of embedding audio and video files on your pages.

FLASH

Macromedia Flash is a powerful software program for creating interactive Web site components. In some cases sites are created entirely in Flash; in others Flash movies and animations are added to HTML-based sites. In this sense Flash goes beyond the programs we've mentioned thus far, such as QuickTime and Windows Media Player, which are primarily media players with some file creation capabilities. Although complete coverage of Flash is far beyond the scope of this book, in this section I'll give you a brief overview of the program and the way Flash files can be placed on your Web site.

Flash Components

Flash includes a variety of tools and capabilities. It is primarily a vector-based drawing and animation program. Vector graphics use geometric formulas, unlike bitmapped raster images, which are composed of dots. Vector images are both compact in file size and scalable, meaning they can be easily reduced and enlarged. (Bitmapped images cannot be enlarged without considerable distortion.) The Flash animation program uses vectors as well, resulting in smaller file sizes and smoother animation.

Flash files can contain other content beyond vector images. Flash can also handle bitmapped graphics, as well as sound and video, and the program includes its own scripting language, ActionScript, which is similar to JavaScript. ActionScript can be used to create interactive sequences, including forms to collect information from users. Flash MX, the newest version of Flash, includes an advanced video compressor (also built into the most recent Flash Player), creating extremely small video files.

Flash files are referred to as Flash movies and can be played only by the Flash Player. Although QuickTime, RealNetworks, or Windows Media files can be inserted into Flash movies, users do not have to have separate players installed to view them. The Flash Player can render any content included in a Flash movie.

Flash movies are built around a series of frames, each one representing a still picture. Flash authors can control the speed, duration, and order of these frames within the animation. Flash also includes a tool called tweening, which allows authors to create the beginning and the end frames in a sequence and then apply a motion or shape tween to create the frames in between the two.

Flash Advantages

Many designers have been intrigued by the power and creative possibilities of Flash; it's no wonder that you see Flash animations throughout the Web. Flash can produce high-quality animation with relatively small file sizes, and it can integrate a variety of other file formats, including audio and video. The ActionScript language is also a powerful tool for creating interaction and sophisticated presentation of information. Moreover, Flash software provides greater control over layout and design than HTML and CSS, which can be influenced more directly by the user and the browser.

In addition, the Flash Player is available for all major browsers and platforms. It can run files in a variety of formats, including QuickTime and Windows Media, as long as those files have been embedded in a Flash movie. Flash movies can also be run in stand-alone applications, using a presentation format called Projector, which can be embedded in the movie itself. Thus, Flash movies can be shared on DVDs, CD-ROMs, or other storage devices rather than being mounted on the Web.

Flash Disadvantages

Even though Flash offers a variety of advanced capabilities, it is not a substitute for HTML, and in fact, there are many situations in which Flash is an inappropriate medium. Flash movies always require a plug-in, which is available in most Web browsers, but the most recent versions of Flash require the most recent Flash Player as well. Although the player is relatively small (383K for Windows, 900K for Macintosh OS X), it still must be downloaded and installed in the user's browser. Moreover, older browser versions may not be able to handle some of the Flash features. And if the Flash Player is not installed, browsers do not automatically redirect the user to an HTML version of the site. Flash authors must create detection utilities within their Flash movies in order to create a redirect.

In addition, search engines have always had problems with Flash. In fact, if you want your Flash site to be indexed by search engines, you'll need to create alternate HTML content that can be searched. Flash MX and Flash Player 6 have increased accessibility features, but currently these features have not been widely implemented in browsers. In the past the lack of accessibility has been one of the major complaints about Flash sites, and more assistive technology is still available for HTML-based sites.

Sites that are largely text based—that is, those that rely mainly on textual information with simple graphic support—are simply not suited to Flash. It is difficult to select and print text content from a Flash file, although printing for Flash pages can be enabled using ActionScript. In the case of simple, text-oriented sites, it's quicker and easier to use HTML than to go to the expense of creating an entire Flash site.

Furthermore, Flash comes with a considerable learning curve. Although you can learn basic HTML and CSS fairly easily, learning Flash requires a considerable investment of time and effort, even to produce a simple Flash-based site. Nonetheless, the rewards to learning Flash can be considerable.

Flash File Types

When you create a Flash movie, you work with what's referred to as a Flash document, which has the extension .fla. This document is created and saved within Flash but is not uploaded to the Web. When authors "publish" a Flash movie to the Web, the program creates a new movie file that has an .swf extension (i.e., a Shockwave file, for Shockwave, which was an earlier version of Flash). The .swf file is an optimized version of the original Flash document; much of the information in the original .fla document is discarded during the conversion, which means it's always a good idea to keep a backup copy of an .fla file. Although the layers in a Flash document are retained, other elements are flattened, and unused elements are discarded. Other elements are optimized according to their original file formats.

It's the `.swf` file that you'll upload to the Web and integrate into your HTML page. Flash contains a utility that will generate the HTML code, but you may want to either create the code yourself or modify what the utility produces.

Embedding a Flash File

Like the developers at QuickTime, Macromedia recommends using the `object-plus-embed` technique described earlier to embed Flash files on your page. There are some slightly different `param` names for Flash files, however. Some common parameters for the Flash Player are described in Table 12–3.

Here's an example of the `object-and-embed` combination for the Flash player:

```
<object classid="clsid:D27CDB6E-AE6D-11cf-
96B8-444553540000"
    codebase="http://download.macromedia.com/pub/
    shockwave/cabs/flash/swflash.cab#version=6,0,0,0"
```

TABLE 12–3   Flash Player Parameters

PARAMETER	DESCRIPTION	FORMAT
play	Determines whether file starts playing automatically when loaded	`<param name="play" value="true" />` or `<param name="play" value="false" />`
menu	Determines whether Flash Player menu provides complete controls or limited controls (limited controls are more common)	`<param name="menu" value="true" />` or `<param name="menu" value="false" />`
quality	Controls how the movie's artwork is rendered in the browser (most common value is `autohigh`)	`<param name="quality" value="low" />` or `<param name="quality" value="autolow" />` or `<param name="quality" value="autohigh" />` or `<param name="quality" value="high" />` or `<param name="quality" value="best" />`
scale	Controls how movie is scaled in window, defined by width and height attributes (default is `showall`)	`<param name="scale" value="showall" />` or `<param name="scale" value="noborder" />` or `<param name="scale" value="exactfit" />` or `<param name="scale" value="noscale" />`

```
        width="400" height="300" id="movie">
        <param name="movie" value="movie.swf">
            <param name="quality" value="high">
        <embed src="movie.swf" quality="high"
            width="400" height="300" name="movie" align=" "
            type="application/x-shockwave-flash"
        pluginspage="http://www.macromedia.com/go/
        getflashplayer">
</object>
```

CHAPTER SUMMARY

Multimedia Basics

❑ Multimedia pages can use various file types, including text, graphics, sound, video, and animation.

❑ Sound, video, and animation files can be quite large and slow to download. They should be used only when they can add necessary information to a Web site.

❑ Users should be given a choice about whether to download a multimedia file.

❑ Web browsers use helper applications called plug-ins to play files other than text or graphics. The browser identifies the necessary plug-in by the MIME type assigned to the file. If the correct plug-in is not already installed in the browser, the user may have to download the plug-in.

❑ As with graphics and text, designers must be careful not to violate the copyright of a creator of a media files. Designers must obtain permission for use of any media file unless it has been offered for promotional purposes or is part of a royalty-free collection.

File Formats and Players

❑ Many formats for sound, video, and media players are available for Web browsers, but not all of them are suitable for all applications.

❑ Sound file formats include `.au`, `.aif`, `.mp3`, `.mid`, `.ra`, `.wav`, and `.wma`. `.mp3` files are popular for digital music files, whereas `.wav` files are popular for sound effects.

❑ Video formats include `.avi`, `.mpg`, `.mov`, `.ram`, and `.wmv`. Video files should be used in the smallest version possible.

❑ Popular media players include QuickTime, RealPlayer, Windows Media Player, and Flash Player.

Placing Files on Pages

❑ You can link to media files, or you can embed them on your pages, depending on the degree of integration you want and the degree of control you want to give your users.

❏ Linking to a media file uses the same code used to link to a page. The file will be played on whatever media player is available on the user's system.

❏ An embedded media file actually plays on the Web page rather than on a separate page.

❏ The code for embedding media files is in a confused state. The W3C specifies the `object` tag for embedding media files in XHTML; the older `embed` tag has been deprecated.

`object` and `embed`

❏ Since Internet Explorer does not support the `object` tag specified by the W3C, many Web designers use a combination of the `object` tag with ActiveX information for Internet Explorer and the `embed` tag for all other browsers.

❏ The `object` tag is used with a `type` attribute to supply the MIME type and the `data` attribute to supply the file name. If the `height` and `width` attributes are given percentage values, they refer to how much of the browser window the object will take up. If they're given in pixels, they refer to the exact width and height of the object.

❏ Designers should supply text between `<object>` and `</object>` that will be displayed if the browser can't play the file itself.

❏ Internet Explorer requires `classid` and `codebase` attributes to supply information about ActiveX elements. However, other browsers will ignore any `object` tag that includes `classid` and `codebase`.

❏ `Param` tags can also be included with `object` tags to supply additional parameter information needed to run particular files.

❏ `Param` takes several attributes, most frequently `name` and `value`.

Flash Basics

❏ Macromedia Flash is a powerful software program for creating interactive Web site components.

❏ Flash is primarily a vector-based drawing and animation program. However, Flash movies can also contain bitmapped graphics, sound, and video.

❏ Flash movies are built around frames that represent still pictures. Authors control the speed and order of the frames and can use tweening to create animation sequences between first and last frames.

❏ Flash can produce high-quality animation with relatively small file sizes, while integrating a variety of other file formats, including audio and video.

❏ Flash is not suitable for all situations. The Flash Player must be installed in the correct version to play Flash movies, search engines cannot index Flash files, Flash has some accessibility problems, text-based sites are not appropriate for Flash technology, and Flash requires considerable training to be used effectively.

Embedding Flash Files

❑ Flash has two file types: `.fla` files are worked on within the program (i.e., off-line), and `.swf` files are uploaded to the Web.

❑ The Flash Player has its own set of parameters to be used with the `object` tag.

❑ Flash movies are embedded using the `object-plus-embed` method.

HTML/XHTML TAGS IN THIS CHAPTER

TAG	EFFECT
`<a></a>`	Links sound or video file to a Web page
`<object></object>`	Embeds sound, video, or Flash file within a Web page
`<param>` (HTML) or `<param />` (XHTML)	Adds additional information necessary for playing a media file
`<embed></embed>`	Embeds sound, video, or Flash file within a Web page; embedded within `<object></object>`; deprecated in XHTML

HTML/XHTML ATTRIBUTES IN THIS CHAPTER

ATTRIBUTE	VALUE	EFFECT
`href`	URL	Hypertext reference; indicates URL of link
`data`	File name	Gives the name of the file being embedded; used with `object`
`width`	Percentage or pixel number	Gives the width the object should take up when it's played; percentage refers to width of browser window
`height`	Percentage or pixel number	Gives the height the object should take up when it's played; percentage refers to height of browser window
`type`	MIME type	Gives MIME type of file
`classid`	Identifying number for ActiveX control or Java applet	Required for Internet Explorer for Windows; ignored by other browsers; used with `object`
`codebase`	URL	Provides location for plug-in; used with `object`
`name`	With `object`, creates unique name for object	With `object`, principally used for scripting

Continued

Continued

ATTRIBUTE	VALUE	EFFECT
	With `param`, identifies parameter being described (e.g., `auto-start, controller`); exact value depends on media player	With `param`, designates the parameter being described
value	Depends on parameter being described	Sets value for parameter being described (i.e., `true, false`, etc.)
valuetype	`data, ref, object`	Sets type for parameter's value
pluginspace	URL	Gives location of plug-in; used with `embed`
src	File name	Gives name of media file with QuickTime; also used with `embed` and Windows Media Player
autostart	`true or false`	Determines whether file begins playing automatically; used with `embed` and Windows Media Player
showcontrols	`true or false`	Determines whether player controls are visible; used with `embed` and Windows Media Player
autoplay	`true or false`	Determines whether file begins playing automatically; used with `embed` and QuickTime
controller	`true or false`	Determines whether player controls are visible; used with `embed` and QuickTime

CLASSIDS IN THIS CHAPTER

CLASSID	MEDIA PLAYER
`CLSID:22D6F312-B0F6-11D0-94AB-0080C74C7E95`	Windows Media Player
`CLSID:02BF25D5-8C17-4B23-BC80-D3488ABDDC6B`	QuickTime
`CLSID:D27CDB6E-AE6D-11CF-96B8-444553540000`	Flash Player

CODEBASES IN THIS CHAPTER

CODEBASE	MEDIA PLAYER
`http://activex.microsoft.com/activex/cont-rols/mplayer/en/nsmp2inf.cab#Version=6,4,5,715`	Windows Media Player
`http://www.apple.com/qtactivex/qtplugin.cab`	QuickTime

Continued

Continued

CODEBASE	MEDIA PLAYER
`http://download.macromedia.com/pub/shockwave/cabs/flash/swflash.cab#version=6,0,0,0`	Flash Player

PLUGINSPACES IN THIS CHAPTER

PLUGINSPACE	MEDIA PLAYER
`http://www.microsoft.com/windows/mediaplayer/`	Windows Media Player
`http://www.apple.com/quicktime/download/`	QuickTime
`http://www.macromedia.com/go/getflashplayer`	Flash Player

MIME TYPES IN THIS CHAPTER

MIME TYPE	PROGRAM
`audio/basic`	Basic audio (`.au`)
`audio/x-aiff`	Audio interchange file format (`.aif`)
`audio/mpeg`	Moving Picture Experts Group Layer-3 Audio (`.mp3`)
`audio/mid`	Musical instrument digital interface (`.mid`)
`audio/x-pn-realaudio`	RealAudio (`.ra`)
`audio/x-wav`	Waveform Audio File Format (`.wav`)
`audio/x-ms-wma`	Windows Media Audio (`.wma`)
`video/x-msvideo`	Audio Video Interleave (`.avi`)
`video/mpeg`	Moving Picture Expert Group (`.mpg`)
`video/quicktime`	QuickTime (`.mov`)
`video/vnd.rn-realvideo`	RealVideo (`.ram`)
`video/x-ms-wmv`	Windows Media Video (`.wmv`)
`video/x-ms-asf`	Windows Media Video active streaming file (`.asf`)
`application/x-shockwave-flash`	Flash (`.swf`)

SOURCES

Castro, E. 2003. *HTML for the World Wide Web*, 5th ed. Berkeley, CA: Peachpit Press.

Gosselin, D. 2004. *XHTML: Comprehensive.* Boston: Thomson Course Technology.

The Future: XML

XML AND THE WEB

In some ways it's misleading to refer to XML as the future, implying that it's somewhere down the road. In fact, XML is already a major component in a variety of applications. It is either in use in or in plans for most modern software, particularly in database and content management programs. Document design programs, such as QuarkXpress and Adobe InDesign, produce XML-coded documents, and Microsoft has announced that all future releases of its Office suite of programs will produce files formatted with XML specifications. As Jeffrey Zeldman (2003) points out, "XML combines standardization with extensibility (the power to customize), transformability (the power to convert data from one format to another), and relatively seamless data exchange between one XML application or XML-aware software product and another" (106).

You've already encountered one XML application: XHTML is the version of HTML developed as an XML language. In fact, XHTML may be the only XML application you actually write as a Web designer, but it helps to know how XML works and what kinds of things XML can be used for. In a single chapter, we can't cover all of the rules and procedures for creating XML documents, but this chapter will give you an overview. We'll discuss the background of XML and will then describe the way XML applications are set up. Finally, we'll discuss some existing XML applications and the way you can expect to interact with XML in the future.

WHY XML?

The original goals for HTML reflected the relatively modest goals for the Web itself. According to Tim Berners-Lee (1999), HTML "was to be a simple hypertext language that would be able to provide basic hypertext navigation, menus, and simple documentation such as help files, the minutes of meetings and mail messages—in short, 95 percent of daily life for most people" (41).

The modest structure and vocabulary of HTML became problematic as the Web expanded into the multifaceted information structure it represents today. HTML was designed to indicate the structure of a document, that is, its major parts. In other words, HTML is focused on display of information, but there is no way in HTML to convey the nature of the information that the document contains. Thus, HTML has tags to indicate what constitutes a paragraph, but it has no means for indicating what that paragraph is about. As the Web developed more database-backed sites, offering everything from online catalogs to libraries of documents, it became more vital to find ways of locating and presenting information within vast collections and of customizing that information to meet the needs of users.

In response to the need for a more comprehensive kind of code, in the mid-1990s the W3C began working on a markup language that would, in Erik Ray's words (2001), "combine the flexibility of SGML with the simplicity of HTML" (11). SGML (standard generalized markup language) had been developed in the 1970s as a text-description language that could be used to encode documents for use in a number of different applications. It has been used extensively in industry and in government agencies both in North America and Europe, but its complexity has made it difficult to adapt as a Web language. HTML was developed as a simplified sublanguage of SGML.

XML is also a sublanguage of SGML, but it has a more far-ranging set of goals. Using XML, you can mark up any kind of data so that it can be identified and processed by a computer. Even though XML stands for "extensible markup language," it's somewhat misleading to think of it as a language at all. In fact, XML is a set of rules and standards for creating markup languages. Using XML, you can develop a syntax for the particular kind of information your organization produces, and you can make it easier for users to locate and customize that information for their own purposes.

The trade-off is that, as you'll see soon, the rules for XML are quite strict. Although parsers (i.e., the programs that interpret code) are able to tolerate some flexibility in HTML code, they can tolerate much less with XML. In fact, the XML specification states that if the XML code contains fatal errors, the XML parser is not allowed to continue processing the code. And a fatal error is any error that keeps the XML from being well formed, that is, meeting the syntactic rules contained in the XML 1.0 specification.

XML BASICS

XML and HTML may look similar on the surface; like HTML, XML has tags, attributes, and values. But HTML creates Web pages, whereas XML is a language for creating other languages. Designers can use XML to create their own custom markup languages and then use those new languages to format their documents, using the new tags to describe the information that the documents contain. Elizabeth Castro (2001) points out that when a tag identifies data, those data then become available for other tasks (13). The labeled information can be located, manipulated, and presented in a variety of software

SIDEBAR: XML TERMS

Application: A customized XML markup language with tags that describe the information they contain.

Document type definition (DTD): One way of declaring a schema for an XML application, using an existing SGML DTD format.

Element: An XML start tag and end tag, along with any information contained between them.

Empty element: An element with no content between start and end tags; often written as a single tag—for example, `<br />`.

Parser: Program that analyzes the grammatical structure of code and then transforms it into a data structure that can be used by another program, for example, a Web browser.

PCDATA: Parsed character data; the content (i.e., data) between the tags in XML elements.

Schema: Description of the structure and content (e.g., elements and attributes) that are allowed in a particular XML application, above and beyond the basic requirements for well-formed XML.

XLink: XML technology that creates links between XML documents.

XML schema definition (XSD): A newer and more extensive way of declaring the schema for an XML application, using XML syntax; supersedes DTDs.

XPath: An XML technology that's used to search through XML documents and locate particular pieces of information needed by other technologies, such as XSLT.

XPointer: XML technology that can be used with XLink to locate particular pieces of information within an XML document.

XSL, XSLT: Extensible style sheet language and extensible style sheet language transformations; allow users to take content from one XML document and reuse it in other XML documents.

---

and Web-based applications. These new, customized languages created using XML are called **XML applications.**

Element has a particular meaning in XML: it's the information from a start tag to an end tag, along with everything in between those two tags. The following XHTML code constitutes an XML element:

```
<p>
My first paragraph.
</p>
```

Here we're using an XHTML tag that's already familiar, but the power of XML comes in its ability to create new tags, so that you can label your content with tags that are meaningful in a particular context. Let's say you were working on a database of tropical fish. You might want to develop tags for the common name, the scientific name, and the area where the fish is commonly found, among other things. Your XML might look like this:

```
<fish>
<commonname>
Blue Tang
</commonname>
<scientificname>
Acanthurus coeruleus
</scientificname>
<location>
West Indies
</location>
</fish>
```

Here we have an element called `fish`, which contains other elements: `commonname`, `scientificname`, and `location`. These elements should look familiar to you since they're very similar to HTML tags. They can contain almost anything, including text and other elements; the name of the element usually identifies the content. As with HTML, the start and end tags use angle brackets, and the end tag begins with a forwardslash.

XML can also use attributes to supply more information about the elements. For example, we could have the following:

```
<name type="common">Blue Tang</name>
<name type="scientific">Acanthurus coeruleus</name>
```

As with HTML, attributes are included in the start tag, with quotation marks for the values. Attributes are considered metadata, that is, information about the data in the element rather than additional content. Thus, `common` in the last example indicates the kind of name being included in the element, but the name itself wouldn't be placed as an attribute.

Requirements for Well-Formed XML

XML has definite requirements that must always be followed in creating elements and attributes if the XML code is to be well-formed, and only well-formed code can be processed by an XML parser. Most of these requirements are already familiar from XHTML.

Every Start Tag Must Have an End Tag or Be an Empty Element. In earlier versions of HTML, some tags, like `<p>`, did not require close tags. Browsers would end the paragraph when a new block-level tag was used. However, XML parsers require all elements to have close tags. Because they don't have to handle various exceptions to XML rules, XML parsers can be much simpler and also much more standardized than HTML parsers.

An **empty element** has no content between the start and end tags. You've already seen empty elements in XHTML, including `<br />` and `<link />`. Since XML requires all tags to have end tags, these empty elements must have either end tags or their equivalents. The single tag version of the start/end tag ends with `/>` (e.g., `<br />`}, but you could also have `<br></br>`—it would mean the same thing. The space before the slash in the single tag version isn't part of the XML specification; it's used to keep older browsers from misreading the tag.

Tags Cannot Overlap. As you saw with HTML and XHTML, tags can be nested inside one another when more than one tag is applied to the same content. That's also true with XML, but the nested tags cannot overlap because overlapping tags do not nest properly. The following tags are correctly nested:

```
<li><a href="page.html">Link</a></li>
```

The following tags overlap (i.e., they're not correctly nested):

```
<li><a href="page.html">Link</li></a>
```

The rule of thumb is that tags must close in the order in which they open. Thus the tag that opens first, closes last. Again, HTML parsers could sometimes deal with overlapping tags, but XML parsers can't.

XML Documents Must Have One Root Element. All XML documents require a root element containing all the other elements in the document. In the earlier example, the root element is `<fish>`; it contains everything else in the document. Only comments and processing instructions (e.g., the opening declaration of the XML version) are allowed to be placed outside the root element.

Element Names Must Follow XML Naming Conventions. Like most computer languages, XML tags must follow certain conventions. The tag name must start with a letter or an underscore character (_); numbers and other characters can be used after the first character. Further, the tag name cannot have any spaces, and names can't include colons because colons are used in namespaces in XML. In addition, names can't include the letter combination `xml` either in upper- or lowercase. And finally, there can be no space before the tag name although there can be space after it (i.e., `<br />`).

XML Is Case Sensitive. Although HTML is not case sensitive, XML is. In XML `<p>` and `<P>` are different tags referring to different things. In fact, XHTML uses only lowercase tags to get around this problem with cases. Given this case sensitivity, it's a good idea to pick a naming convention and stick to it. For example, you could use `commonName`, `commonname`, or `Commonname;` but once you've decided on a capitalization method, you'd use the same one on `scientific name` and all other tags. Keeping the convention consistent will help you write well-formed XML.

Values Must Be Inside Quotation Marks. Like HTML, XML can include attributes and values. In HTML those values don't always have to be in quotation marks;

in XML they do. You can use either single or double quotation marks, but you must use them in pairs (i.e., `attribute="value"` or `attribute='value'`).

Special Characters Must Be Encoded or Declared in a DOCTYPE (DTD) or XML Schema (XSD). HTML uses ASCII code to represent several special characters ("character entities") like the ampersand (see appendix 5 for more on character entities). XML allows only these five characters:

```
&amp; for ampersand, &
&lt; for less than sign, <
&gt; for greater than sign, >
&quot; for quotation mark, "
&apos; for apostrophe, '
```

Any other special characters must be defined in the DTD or XSD.

XML also handles white space differently than HTML does, although this doesn't necessarily constitute a requirement. HTML browsers strip out any extra white space when the document is processed. Thus, if you put double spaces after the periods in your sentences, they'll be converted to single spaces by the browser. However, according to the XML specification, this white space won't be stripped out by XML parsers as long as the space occurs in PCDATA. Currently, though, Web browsers still strip the white space from XML, as they did from HTML.

XML DOCUMENTS

Even from this brief description, you can see the possibilities that XML offers. You can create your own markup to present the particular information your organization deals with and make it available to other organizations who can, in turn, use your new language to mark up their own information if they choose.

XML documents have two sections, sometimes called the prolog and the body. The body section includes the actual XML elements that you've created, along with the data they enclose. The prolog, similar to the head section in an XHTML document, includes the XML declaration, processing instructions, and document type definitions (DTDs) or XML schema definitions (XSDs). The prolog is not required for XML documents, but it includes many things that can prove helpful for both parsers and human readers. A sample prolog (with DTD) looks like this:

```
<?xml version="1.0" encoding="UTF-8"
    standalone="no"?>
<!DOCTYPE fish SYSTEM "http://www.tropicalworld.com/
    tropicalfish.dtd">
```

XML Declaration

Each XML document begins with a **declaration** that indicates the XML version being used. The question marks at the beginning and end of the declaration are processing instructions to the parser. The version attribute is required if you're including an XML declaration, but right now the only version of XML is 1.0, so the value is always the same. (A version 1.1 was issued in 2004, but its features are mainly used by mainframe programmers.)

Two other attributes, although not required, are frequently used in the declaration. Encoding refers to the character encoding being used in the document, most frequently `encoding="UTF-8"` or `encoding="UTF-16"`. Unicode (i.e., UTF) is a system that enables characters from a variety of languages to be used; and since XML is an international format, the ability to use UTF characters is an important advantage.

`standalone` is an attribute that can be used to indicate whether the document needs other files to be read. If the document exists on its own, without reference to other files, the attribute and value are `standalone="yes"`; if the document uses an external DTD or other file (e.g., an external CSS stylesheet), the attribute and value are `standalone="no"`.

Schemas

When you create your XML application, you must specify which elements and attributes are allowed and required in your new markup language so that the parser can judge whether documents that are written using the application are valid. A document is considered valid if its data conform to the set of content rules that you define, describing the acceptable data values and format. These content rules are called a **schema.**

Schemas describe XML application languages, explaining any constraints placed on the structure and content of documents above and beyond the basic constraints of XML. These schemas are used to keep documents consistent that use the same application. All documents that correctly utilize a schema will be considered valid, meaning that their data are in the right form.

Document Type Definitions (DTDs). There are two ways of creating schemas: document type definitions (DTDs) and XML schema definitions (XSDs). DTDs are an older way of creating a schema, but they're still widely used in a variety of XML applications. In fact, you've already encountered one use of the DTD with XHMTL. Like CSS a DTD can be either internal or external.

Internal DTDs are used most frequently for XML applications that apply to a single document. The DTD declares the elements and attributes that are part of the particular XML application for the document. A complete DTD includes everything that's part of the rules for the particular XML application.

External DTDs are used for sets of related documents that will all use the same DTD. Rather than placing the same internal DTD on every document in the set, you can create an external file with the DTD and then place a reference to that file at the beginning of all the documents that use it. You begin by creating the DTD in a separate text file in which you define the rules for the DTD, saving it with the `.dtd` extension.

You can link to DTDs that you've created yourself for your own documents, or you can link to one that was created by someone else—a public external DTD. If you create your own DTD, you'll link to it by providing the URL for the DTD, as

SIDEBAR: DEFINITION VERSUS DECLARATION

Although I make reference to the document type definition, or DTD, throughout this section, you'll also encounter references to the document type declaration. Although these two terms may seem similar, they're not the same thing. The document type definition refers to the actual set of rules that set forth the elements and attributes of an XML application. The document type declaration, on the other hand, includes all of the code associated with the definitions, beginning with <!DOCTYPE.

you did with your external CSS, along with the word *SYSTEM* to indicate that it's a customized DTD (I used that format for the DTD in the prolog example earlier). With a public external DTD, you'll also include the name of the DTD with the word PUBLIC to indicate that it's a publicly available set of XML rules.

XML Schema Definitions (XSDs). DTDs worked effectively while XML was in its initial stages, but they do have some limitations. DTD syntax, for example, is somewhat different from XML syntax because it was based on earlier SGML DTD syntax. Unfortunately, the syntax differences have meant that newer XML tools and standards have trouble with DTDs. In addition, with DTDs two elements can't use the same name, even if they're used in different contexts. But most importantly, DTDs cannot indicate the type of information that must be included in a given element or attribute, only its syntax. Thus, users can't stipulate particular types of information that must be included for a document to be considered valid.

In contrast, XML schema definitions are written in XML rather than SGML. This format allows for both global elements (i.e., elements that are used in the same way throughout a document) and local elements (i.e., elements that have different meanings in different contexts). XML schema definitions introduced a system of data types, which allow authors to specify exactly what kind of data an element should contain. Thus, you can indicate whether an element should contain a text string or an integer or even a five-character zip code. Schemas also use a considerably more complex syntax in their declaration than a DTD. For example, in a DTD you can indicate that an element contains text by specifying that its content be PCDATA, but in XSD you can go on to specify exactly what kind of text you want the element to contain by using one of several predefined simple types, such as date and integer.

Whether you use a DTD or an XML schema will depend on the kind of information you're dealing with and the kind of applications it will be used for, but both formats are available for XML writers. Although DTDs are still widespread, they will probably be replaced by XSDs over time.

Namespaces

The **namespace** is one more XML component that can make data more usable in XML documents. When you create your own tags, there's a possibility that you may someday need to use the same tag in a different document with a different meaning. This can lead to real problems if you need to combine the two documents. For example, let's say the `fish` XML document in the earlier example was created by a pet store that decided to use the `name` element with the attributes `common` and `scientific`. Later on, the owners want to create a new XML document that lists their customers and the fish those customers have bought, combining the fish database with a customer database. But the XML document for the customers uses `name` to refer to a customer's name. How can the XML authors use the element `name` in two different contexts when they create the new document that lists the customers and their fish purchases?

The creators of XML foresaw this problem. As Bert Bos (2001) points out,

> XML allows you to define a new document format by combining and reusing other formats. Since two formats developed independently may have elements and attributes with the same name, care must be taken when combining those formats. . . . To eliminate name confusion when combining formats, XML provides a namespace mechanism.

Namespaces are labels that distinguish one set of elements from another. This label is in the format of a URL, although the URL needn't point to a specific page. The rationale for using a URL to identify the source of a namespace is based on the fact that URLs are unique (i.e., a domain name can be used for only one site) and permanent (i.e., a domain name doesn't change frequently).

You've already used a namespace in your XHTML code; the opening root element of an XHTML page is actually an XML namespace.

```
<html xmlns="http://www.w3.org/1999/xhtml">
```

In this case the XHTML namespace doesn't refer to an actual page on the W3C site; it simply indicates that XHTML elements are part of the XHTML DTD. XML documents that combine elements from more than one namespace have a variety of namespace formats available, including prefixes and colon delimited names.

Namespaces can also be declared within the document rather than at the document's beginning. So the `name` element could be identified with the `fish` element, using one namespace, and then could use another namespace to identify with the `customer` element. The format for this contextual use is somewhat complex. For now it's enough to know that namespaces exist and function to keep each set of elements and attributes separate and unique.

THE XML FAMILY

XML is a language for creating languages, but it is also a family of technologies that provide services for XML documents. This is an aspect of XML that is continuing to evolve. I'll summarize some of the more widely used XML programs here, but for a more complete overview, you can consult the W3C Web site.

XSL and XSLT

XSL (extensible style sheet language) is an XML language that can create style sheets to be used with XSLT (extensible style sheet language transformations). XSLT allows you to take the content from XML documents and transform it into other documents or formats. The XSL style sheets define the layout and indicate where the data are to come from within the source document. XSL style sheets are XML documents, which means they must follow the rules for well-formed XML.

The process used by XSL is similar to that used by CSS to format HTML, but the XSL style sheet goes far beyond CSS. The XSL style sheet contains a kind of supercode that describes the ways in which the data from the XML document are to be transformed. With XSLT the data can be translated into XHTML for display on a Web site. Or they can be transformed into text for display in a document or into another XML structure to be integrated into another XML application. XSLT uses templates that specify what kind of data to look for in the source document (called a source tree in XSLT) and then specify how to place that data into the new document (called a result tree). The template specifies patterns that the XSLT engine looks for in the source and transfers to the result tree.

Using XSL and XSLT, users can take XML documents provided on Web sites, intranets, and extranets and use the data in new XML applications without having to go through the documents manually to remove the necessary content. As long as the XSLT is written so that it can find the necessary data in the XML document, the content can be reused, reformatted, and re-presented with a minimum of effort.

XPath

XPath is an important tool used in XSLT and other technologies that work with XML documents. It's similar to languages like SQL, which are used to locate information in databases. XPath is the language used by XML technologies to find information in XML documents. In other words, XPath is the language that XSLT uses to look for patterns in a source document. XPath then returns that information to XSLT so that XSLT can perform some kind of operation on the information. By using XPath, technologies like XSLT can select particular pieces of information from XML documents without processing information that isn't needed.

XLink

Links in XML offer more possibilities and far more complexity than links in HTML, resulting in a separate XML technology for linking. XLink, or XML linking language, is a relatively new technology that has not been as widely implemented as XSLT. However, its use should increase as XML becomes more widespread.

Since XML is intended only to describe data, it contains no tools for linking XML documents to one another or to other online resources. XLink was created to provide this capability. It provides the same linking power that the anchor element provides in XHTML, but it also provides more. Any element in an XML document can perform as a link if the XLink namespace is added to the root element tag as an attribute. With XLink extended links, two or more resources can be connected using a single link, and that link need not be contained within any of them. For example, if you were creating a database of endangered species and wanted to bring together an XML document with a list of tropical fish and an XML document about their status on the endangered species list, an extended link would allow you to link both documents to your third document on endangered species. XLink also allows you to add instructions that specify how links should be treated, for example, whether they should be opened in new windows or the same window or should be embedded in the current document, like multimedia files.

XPointer

XPointer actually works with XLink, using the XPath language to perform similar tasks. Whereas XLink links to a complete document, XPointer can be used to find a specific piece of information within that document. XPointer uses the on-page-anchor concept that we discussed in chapter 8. Like XLink, XPointer attributes can be added to any XML element, attributes, or combination of the two. Like XPath, XPointer locates nodes of information, but it also allows you to locate other parts of a document beyond nodes. By using a combination of XLink and XPointer, writers can link to entire documents or smaller pieces of information, depending on the needs of the XML link being established.

Other XML Technologies

Other XML technologies exist that are more clearly related to programming than to markup. In many cases these technologies are APIs (application programming interfaces), sets of definitions or procedures that describe the way software programs communicate with each other. Unlike most computer languages, XML is a language for describing data structure rather than a set of functions and procedures. APIs provide the functions and procedures that XML doesn't have built in, such as mathematical calculations. There are two major APIs for XML: DOM (document object model) and SAX (simple API for XML). Both provide additional capabilities for XML documents.

XML-RPC and RDF are ways of exchanging XML data and making procedure calls across the Web. SOAP (simple object access protocol) is a standard for exchanging XML-based messages over a computer network, normally using HTTP; it allows information exchange in a platform-independent network. SMIL (synchronized multimedia integration language) is a W3C recommendation for describing multimedia presentations using XML, enabling nonprogrammers to author multimedia presentations in a text editor. And SVG (scalable vector graphics), which we mentioned in chapter 6, is an XML markup language for describing vector graphics.

Finally, RSS (rich site summary) is an XML technology originally used by Weblogs. It provides short summaries of Web content together with links to the full version. RSS has been widely used by bloggers to share headlines or the full text of entries. Programs called feed readers can check blogs using RSS and display updated articles. Most major Web sites now include RSS feeds. In 2004–2005 use of RSS spread to many major news organizations, including Reuters and the Associated Press, and many news sites now allow other Web sites to incorporate their syndicated headline or headline-and-short-summary feeds. For more about blogging and RSS, see appendix 4.

Again, this is just a brief summary of some current XML applications in use; more are being developed daily. Clearly, XML is changing the way information is located and processed on the Web and elsewhere.

XML AND CSS

Unlike HTML, XML tags have no default formatting at all. Thus, XML writers must specify exactly how they want their content to be displayed. In fact, XML documents won't be displayed in Web browsers unless the information about display styles is provided. XSLT is one way to apply styles. Because XSLT can convert XML to HTML, styles can be applied by using the default formatting built into HTML tags, such as paragraphs, lists, and headings. However, XSLT itself does not provide any formatting instructions.

The more common way to format text in XML is to use CSS. Although CSS can be applied directly to XML, it can also be applied to HTML that is generated by XSLT. However, it does make a difference whether the CSS is applied to XML documents that have no formatting whatsoever or to HTML, which already has default formatting applied.

CSS used with XML is most commonly applied through an external style sheet, although, like HTML, XML documents can also use internal style sheets and style directions that are placed within tags. The same external style sheet can be applied to several XML documents, as it can to several HTML pages. The link for a style sheet in XML is similar to the one you've used for HTML, but there are some differences. The link tag is not used since it's part of HTML; and since this is an XML processing instruction, it begins and ends with ?:

```
<?xml-stylesheet type="text/css" href="styles.css"?>
```

The style rules that are placed on this style sheet will probably involve many more properties and values than those used with HTML elements. Remember that the XML elements have no default styles applied to them. Thus, unless the XML is being converted to HTML, every characteristic must be specified for elements: font, size, color, line height, indentation, margins, spacing, and so on. The element must also be set to be displayed as a block element or an in-line element. XML pages use CSS layout properties, just like HTML pages. However, unless at least some of these style rules are provided, text will be presented in a solid, undifferentiated block.

A style rule for XML looks very much like a style rule for HTML. The selector is the element name you create, but the way in which the properties and values are expressed is the same. A style rule for the common name element I created earlier would look like this:

```
commonname{font:medium/1.5 georgia, times, serif;
    display:block; margin:10px 50px;}
```

REAL-WORLD APPLICATIONS

As I said at the beginning of this chapter, XML has already been adopted for a variety of applications, but perhaps its greatest impact has been on content management systems.

Content Management Systems

The principle of content management has actually been in existence much longer than XML. All organizations produce quantities of information, and these days most of the information is in digital form. Managing digital information involves making it available to all the various audiences that need it and producing it in the optimal form that each audience demands. **Content management systems** are programs that can both accept digital information and produce it on demand for all those within the organization who need access.

A variety of content management systems are currently in use. Enterprise content management systems help organizations control the transactions and customer relations involved in e-commerce. Publication content management systems handle all the various publications (e.g., manuals, articles, help screens) that an organization may produce. Web content management systems are used with large-scale Web sites that require frequent updates and maintenance of database content. Learning content management systems (also called managed learning environments) are familiar to many students and instructors; these software systems manage online courses, or courses with online content. They provide tools such as managed chat rooms, tools for uploading and downloading content, controlled access for class members, and a single, consistent interface.

The need to manage information is particularly important in organizations in which new products require frequent revisions in documentation or even customized documentation for particular models. Sales personnel need to see current specifications; maintenance personnel need up-to-date procedural descriptions and schematic drawings; customers need instructions for operating and maintaining the specific equipment they've purchased. Content management systems can also track warranties to alert customers to necessary replacements and upgrades of equipment.

Modern content management systems must deal with more than text; graphics, video clips, audio, even live teleconferencing can all be part of the mix.

And information created within the organization may come in a wide variety of formats, created with a mixture of programs and tools, and it may be revised frequently. Effective content management involves working with the smallest useful units of information taken from this mix, realigning them, and delivering them to users in a personalized format.

Most content management systems involve some kind of information storage system (i.e., a database) and a means for accessing that information. Since the information may need to be revised regularly, the technology behind the system must be able to accommodate dynamic content. Most effective content management systems allow users to employ a variety of tools so that they can use whatever program is best for the information being created. Thus, a good system is able to adjust to content from word processors, illustration software, CAD/CAM systems, and spreadsheets, as well as video and audio files as necessary.

In order for users to access this information, the content management systems must be able to reduce it to small units. This means being able to tag, index, search, and reuse units of information according to the needs of the users. Units must be indexed by what they contain rather than by how they look or what tool was used to create them. For example, users in one department might need an engineering schematic, along with instructions for a particular maintenance procedure; those in another department might need the same schematic with a technical article describing an advance in design; and those in a third department might need the schematic, along with a set of specifications. The same file would be repurposed in three different ways to accompany three different chunks of text.

Content Management and XML

Some early content management systems used SGML to provide content markup, particularly for document management. SGML provided a means of creating markup languages that would tag content separate from its presentation. However, the complexity of SGML and its style sheet language (DSSSL) made it difficult to use. In fact, DSSSL has never been fully implemented in an information processor.

HTML was used widely as the front end for Web content management systems. Jeffrey Veen (2001) described this "three-tiered architecture" as object-oriented publishing with a back-end database, a middle-ware content management program like Microsoft's Active Server Pages or Macromedia's Cold Fusion, and an interface designed in HTML that delivered the content to a Web browser (221). But HTML had all the deficiencies mentioned earlier: it emphasized presentation over structure and was unable to indicate the content of its largely text-oriented segments. Clearly another solution was necessary.

The XML simplification of SGML was an obvious answer. Because XML was simpler to use and to process than SGML, low-cost tools could be developed using XML. Because XML was developed by the W3C, it had the support of a wide range of companies, including Microsoft and Sun Microsystems, ensuring its compatibility with both Windows and Unix platforms. A combination of XML and Java provides the technology that runs most content management systems today.

In an XML content management system, content is stored in a central repository, where it can be broken down into reusable units, indexed, and maintained with version control. Using XML data structures, users are able to search through the repository, looking for particular content, properties, or structure. Through XLink and XPointer, the repository can also include links and references to related information.

When content is received in the repository, it's checked against the DTD or XSD to make sure that the format is valid. Then it can be broken down into its component parts, based on its markup. The marked-up units of content can then be searched, recombined, referenced, and published at users' requests. When the content is published, it can be reformatted into the form the user needs through XSL transformations.

Because XML can tag content according to its meaning rather than its appearance, users can custom-create documents by pulling only the pieces of information they want out of the content repository, and they can specify the kind of document in which they want to have the information placed. XML allows content creators to indicate the nature of their content and the links it has to other content already in place. It also allows content users to find, format, and publish information at the time it's needed, without having to filter through information that doesn't fit their criteria. Most of all, maintenance and documentation costs can be lowered by keeping information both up-to-date and readily available.

XML AND YOUR WEB SITE

As I said earlier, you may never actually code XML by hand, even though you may use tools that add customized XML to your content. However, parts of XML will undoubtedly affect your Web sites now and in the future. For example, most blogging software now runs on XML. Thus, when you add a blog to your site, you'll be adding XML (for more on blogging, see appendix 4). And in the future you may use scalable vector graphics (SVG) on your site as the format becomes more widespread. In addition, if your Web site interacts with a database, that interaction will probably be governed by XML procedures. Furthermore, if you work with a large-scale site that uses a content management system, you'll undoubtedly be using XML, although it may be concealed behind an interface.

XHTML is currently the standard for Web design, and the combination of XHTML and CSS will continue to replace older Web site design as older browsers are replaced by more modern browsers. XHTML will probably remain the primary means of displaying XML on the Web, since few browsers are capable of displaying XML code itself.

Because XML helps to make software compatible with a variety of platforms and other software programs, it will undoubtedly continue to be used in the software industry, as well as in the expanding market for Web services. And you'll undoubtedly encounter these XML-based applications, although you may not be creating them yourself. XML is the most visible evidence of the W3C's standard of interoperability. As Jeffrey Zeldman (2003) points out, "That is the hallmark of a good standard: that it works, gets a job done, and plays well with other standards" (114).

CHAPTER SUMMARY

XML Background

❏ XML (extensible markup language) is a major component of modern software, particularly in database and content management programs.

❏ HTML was designed to indicate the structure of documents but not their content. As the Web developed database-backed sites, a markup language was needed that could help to locate, present, and customize information.

❏ XML is a sublanguage of SGML (standard generalized markup language); it is a set of rules and standards for creating markup languages.

❏ XML rules are more strict than those for HTML. Thus, XML parsers can be simpler because they have more absolute rules to work with.

❏ Like HTML, XML includes tags, attributes, and values. However, XML is a language that can be used to create other languages with tags that describe the information contained in documents.

XML Applications

❏ Customized markup languages created with XML are called XML applications.

❏ XML applications must meet certain requirements to be considered XML. Applications that meet these requirements are called well formed; only well-formed XML can be processed by an XML parser.

XML Basics

❏ An XML element consists of a start tag, an end tag, and the content between those two tags.

❏ XML start tags can include attributes that provide metadata, that is, information about the data included in the element.

❏ XML has several requirements: start tags must have end tags or be empty elements; tags cannot overlap; XML documents must have one root element; element names must follow naming conventions; XML is case sensitive; attribute values must be inside quotation marks; special characters must be encoded or declared in a DTD or XSD.

❏ XML browsers will not strip out extra white space, as HTML browsers do.

❏ XML documents have two sections, the prolog and the body. The prolog includes the XML declaration, processing instructions, and DTD or XSD, whereas the body section includes the XML elements and data.

XML Document Format

❑ XML documents begin with a declaration indicating the version of XML being used (currently version 1.0).

❑ XML schemas indicate which elements and attributes are allowed and required in the XML markup language being used in the document. Documents that use a schema correctly are considered valid.

❑ XML schemas can be created using either document type definitions (DTDs) or XML schema definitions (XSDs).

❑ DTDs declare elements and attributes that are part of a particular application, either for a single document (an internal DTD) or a group of documents (external DTD). Public external DTDs are publicly posted and can be used by other XML writers.

❑ XSDs are a more complete way to write schema definitions; they can specify exactly what content must be included in an element or attribute.

❑ Namespaces are labels that distinguish one set of elements from another by providing a URL to identify the application from which the element set is taken.

XML Technologies

❑ XSL (extensible style sheet language) is used with XSLT (extensible style sheet language transformations) to take the content from one XML document and transform it into other documents or formats. XSLT can translate data from XML documents into XHTML, into text, or into another XML structure to be integrated into another XML application.

❑ XPath is an XML tool used in XSLT to locate information in XML documents. XPath is used to look for patterns of information and then return the location to the program that needs to perform an operation on the information.

❑ XLink (XML linking language) is the language used to create complex links within XML. XLink can provide extended links that connect two or more resources with a single link and can also provide instructions that specify how links should be treated.

❑ XPointer uses the XPath language to locate specific pieces of information for XLink. It operates somewhat like an on-page anchor in XHTML.

❑ XML also uses a set of APIs (application programming interfaces) to describe the way software programs communicate with each other. The principal APIs for XML are DOM (document object model) and SAX (simple API for XML).

❑ XML-RPC and RDF are ways of exchanging XML data and making procedure calls across the Web. SOAP (simple object access protocol) is a standard for exchanging XML-based messages across a network.

❑ SMIL (synchronized multimedia integration language) is a W3C recommendation for describing multimedia applications using XML.

❑ SVG (scalable vector graphics) is an XML language for describing vector graphics.

❑ RSS (rich site summary) is XML technology that provides short summaries of Web content along with links to a full-length version. Programs called feed readers check blogs using RSS to display updated articles.

❑ Unlike HTML, XML tags have no built-in formatting, so XML authors must specify the presentation they want for their XML documents. CSS can be applied to XML through an external style sheet and must specify all the styles to be applied, including the way the element is to be displayed (i.e., block or inline).

XML and Content Management

❑ XML is widely used in content management systems, which involve producing, repurposing, and accessing digital information.

❑ Effective content management systems include a data storage repository that can accept files created with a variety of programs. The information in the files must be subdivided into small units that can then be tagged, indexed, searched, and reused.

❑ XML was adopted for use with content management systems because it provided the markup capabilities of SGML without the attendant complexity and expense.

❑ In an XML content management system, content is stored in a central repository, where it can be broken down into reusable units, indexed, and maintained with version control.

❑ Users of an XML-based content management system can employ XML data structures to search the repository, looking for particular content, properties, or structure. Through XLink and XPointer, the repository can also include links and references to related information.

❑ With XSLT, units of content can be recombined, formatted, and published at the time documents are requested.

XML and Web Design

❑ Web designers will interact with XML in a variety of ways: through blogging software, SVG, database interactions, and content management systems.

❑ Web designers will also use XHTML as the primary means of displaying XML on the Web.

SOURCES

Berners-Lee, T. 1999. *Weaving the Web*. New York: HarperCollins.

Bos, B. 2001. XML in 10 points. http://www.w3.org/XML/1999/XML-in-10-points. (accessed June 9, 2005).

Castro, E. 2001. *XML for the World Wide Web*. Berkeley, CA: Peachpit Press.

Ray, E. T. 2001. *Learning XML*. Sebastopol, CA: O'Reilly.

Veen, J. 2001. *The art and science of Web design*. Indianapolis, IN: New Riders.

Zeldman, J. 2003. *Designing with Web standards*. Berkeley, CA: New Riders.

Appendix 1
HTML Tables

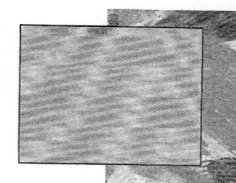

Tables were part of the original HTML specification because the scientists who were the original users of the Web used tables to display data. However, prior to the development and implementation of CSS positioning properties, designers also used tables to create page grids so that they could have more control over the layout of their pages. These layout tables frequently became very complex, with tables nested within tables and transparent single-gif graphics used to create spacing on the page. Not surprisingly, the extra code involved in these elaborate tables created larger page files with increased download time.

Currently, CSS positioning properties provide more efficient control of page layout without the downside of increased file size. Although some older sites using table layout grids are still maintained, it's better not to begin new sites with the complex layout tables favored in the nineties. HTML tables now have returned to their original function of presenting data. For more on CSS positioning, see chapters 10 and 11.

BASIC TABLE LAYOUT

As with print tables, HTML tables are combinations of rows and columns, with table data included in cells within these columns. If you've used spreadsheet programs, you're probably familiar with this basic layout.

HTML tables consist of a series of container tags, with each pair set inside another. The table begins with `<table>` and ends with `</table>`. After the opening `<table>` tag, each row is set up with `<tr></tr>` (i.e., t[able] r[ow]). Within these rows you can use two kinds of cells: table headers (`<th></th>`) and table data (`<td></td>`). You can also use `<caption></caption>` to supply a caption above the table. Table cells are formatted by the browser, but text in header cells is usually boldfaced and centered in the cell, whereas text in data cells is roman and left-justified. The HTML for a simple two-column, two-row table would look like this (I've added a border so that you can see the cells more clearly in the browser):

```
<!DOCTYPE html PUBLIC "-//W3C//DTD XHTML 1.0
    Transitional//EN" "http://www.w3.org/"
    TR/xhtml1/DTD/xhtml1-transitional.dtd">
<html xmlns="http://www.w3.org/1999/xhtml">
<head>
<meta http-equiv="content-type" content="text/html;
    charset=utf-8"/>
    <title>A Simple Table</title>
</head>
<body>
<table border="1">
<caption>The Table Caption</caption>
        <tr>
             <th>
A Header Cell
             </th>
             <th>
A Header Cell
             </th>
    </tr>
             <tr>
             <td>
A data cell
             </td>
             <td>
A data cell
             </td>
        </tr>
</table>
</body>
</html>
```

In a browser it would look like Figure A1–1.

TABLE ATTRIBUTES

Some attributes are available to use with the table tags, but some older attributes have been deprecated from earlier versions of HTML, and some other table attributes are browser-specific. In this section we'll consider only those attributes that are available in the HTML 4.01/XHTML 1.0 specifications.

Border The border attribute for the `<table>` tag allows you to create a visible border around your table cells (the default is no borders at all). For a value, supply a number equal to the width in pixels, but don't use the px abbreviation used in CSS.

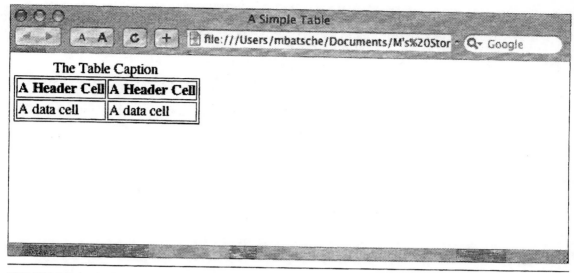

FIGURE A1–1   A simple table

Cellspacing and cellpadding Both cellspacing and cell-padding provide space around the table contents. cellspacing puts space between adjacent cells and along the edges of the tables (the equivalent of the CSS margin property); cellpadding puts space between the contents of a table cell and the edge of the cell (the equivalent of the CSS padding property). The value for both cellspacing and cellpadding is a number equivalent to the number of pixels of space to be added.

Colspan and rowspan Sometimes you want to stretch a single table cell across more than one column or more than one row in your table—for example, when you have a column header that applies to more than one column. The colspan attribute stretches a table cell across two or more columns; the value of the attribute is the number of columns you want the cell to stretch across. rowspan works in the same way to stretch a cell across two or more rows.

Width Browsers automatically draw a table only as wide as necessary to correctly display the cell contents. You can make a table wider than this by using the width attribute. The value is a number of pixels or a percentage that refers to the width of the browser window.

Summary The summary attribute allows you to include a description of the purpose and the content of the table. This summary will be available to screen readers and browsers that cannot display the table itself. Thus, it is another way of making your pages more accessible, and for this reason the summary attribute is required for tables in XHTML.

Abbr Abbr, which stands for *abbreviation*, is another accessibility aid. If a table header or data cell content is too long to be read easily in a screen reader, it can be shortened using abbr.

Here's the same simple table with added attributes:

```
<!DOCTYPE html PUBLIC "-//W3C//DTD XHTML 1.0
    Transitional//EN" "http://www.w3.org/TR/xhtml1/
    DTD/xhtml1-transitional.dtd">
<html xmlns="http://www.w3.org/1999/xhtml">
<head>
    <meta http-equiv="content-type" content="text/html;
    charset="utf-8"/>
    <title>A Simple Table</title>
</head>
<body>
<table summary="A simple demonstration table."
    border="1" width="70%" cellpadding="5">
<tr>
            <th abbr="header">
A Header Cell
            </th>
            <th abbr="header">
A Header Cell
            </th>
            </tr>
            <tr>
            <td>
A data cell
            </td>
            <td>
A data cell
            </td>
        </tr>
</table>
</body>
</html>
```

Figure A1–2 shows what it would look like in a browser.

ADVANCED TABLE TAGS

HTML 4.01 and XHTML 1.0 both provide some new tags that allow you to divide tables into sections and to create running headers and footers. These features are available only in the most recent browser versions, but they provide extra possibilities for data tables. According to the W3C specification, a table that uses `<thead></thead>`, `<tfoot></tfoot>`, and `<tbody></tbody>` can have a body that scrolls while the head and foot sections stay fixed. If the table stretches over

FIGURE A1–2   A table with attributes

more than one page, the table head and foot are repeated on each page that contains table data.

<thead></thead> The <thead> tag allows you to define a set of table rows as a table head. The tag is used only once in a table and is placed just after the table tag. You can place one or more <tr></tr> tags within the <thead> </thead> tags. If you don't add any styles to your <thead>, the browser will display the text centered vertically and horizontally.

<tfoot></tfoot> The <tfoot> tag defines a table footer. Like <thead>, <tfoot> appears once, at the end of the table before </table>. Also like <thead>, <tfoot> can contain one or more <tr></tr> tags. If your table is printed over several pages, the table footer will be repeated at the bottom of each table section.

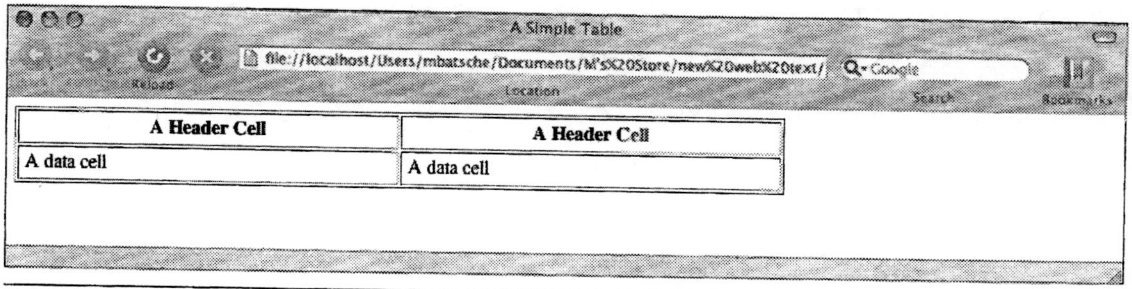

<tbody></tbody> The <tbody> tag divides your table into sections by collecting one or more rows into a group. Unlike <thead> and <tfoot>, <tbody> can be used more than once in a table to group a series of rows together. You can also use CSS classes to apply different styles to these different sections.

rules Using the rules attribute with the table sections allows you to specify the thickness of a cell's internal borders and to specify where the rules are to be drawn (i.e., around the entire cell or only a portion of it). rules have these values:

- ❑ groups—places thicker borders between row and column groups
- ❑ rows or cols—places borders between all rows or between all columns
- ❑ all—draws all borders
- ❑ none—removes all borders

Here's the code for the same simple table, this time with added sections and rules. Note that I've placed a nonbreaking space ( ) in the first cell to provide a blank space.

```
<!DOCTYPE html PUBLIC "-//W3C//DTD XHTML 1.0
    Transitional//EN" "http://www.w3.org/TR/xhtml1/
    DTD/xhtml1-transitional.dtd">
```

```
<html xmlns="http://www.w3.org/1999/xhtml">
<head>
<meta http-equiv="content-type" content="text/html;
    charset=utf-8" />
    <title>Table Sections</title>
</head>
<body>
<table summary="This is a table demonstrating the thead,
    tfoot, and tbody sections." width="75%" cell-
    padding="5" border="5" rules="groups">
    <thead>
    <tr>
        <td colspan="2" rowspan="2">   </td>
            <th colspan="2">
First Column Heading
            </th>
        </tr>
            <tr>
            <td>
First Column Subheading
            </td>
            <td>
Second Column Subheading
            </td>
        </tr>
        </thead>
            <tbody>
            <tr>
            <th rowspan="2">
Row header
            </th>
                            <th>
First Row Subhead
            </th>
            <td>
Table data
            </td>
            <td>
Table data
            </td>
            </tr>
            <tr>
                            <th>
Second Row Subhead
            </th>
            <td>
```

	First Column Heading	
	First Column Subheading	Second Column Subheading
Row header	**First Row Subhead**	Table data	Table data
	Second Row Subhead	Table data	Table data
A footnote to the table.			

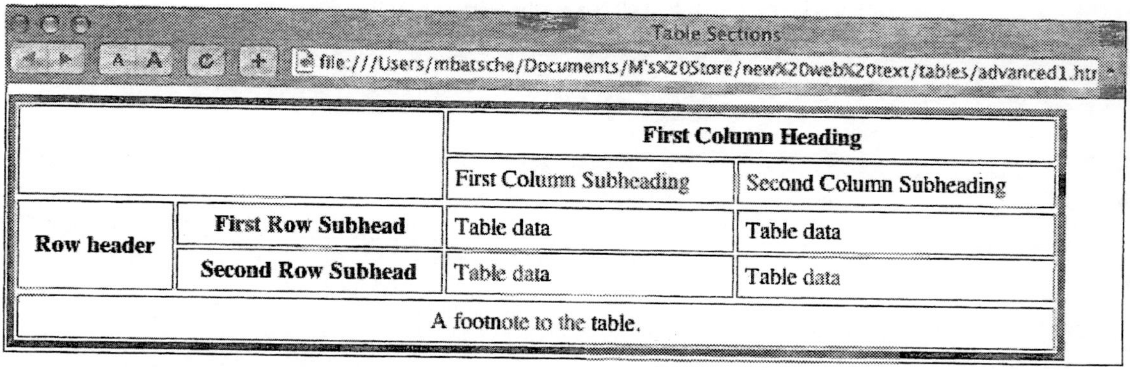

FIGURE A1–3 Table sections

```
Table data
          </td>
          <td>
Table data
          </td>
          </tr>
     </tbody>
        <tfoot>
     <tr>
          <td colspan="4" style="text-align:center;">
A footnote to the table.
          </td>
     </tr>
     </tfoot>
</table>
</body>
</html>
```

In a browser it would look like Figure A1–3.

ADDING STYLES TO TABLES

Earlier versions of HTML included attributes to apply background colors, text colors, text alignments (both horizontal and vertical), border colors, and so on. Most of these attributes have now been replaced by CSS properties. You can use all the table tags as selectors on your style sheets and choose fonts, colors, alignments, and other properties as you can for other HTML elements. An external style sheet for a table might look like this:

```
table {font-family:verdana, helvetica, sans-serif;
border:#00F medium ridge;}
```

```
td {color:#F30; font-size:medium;}
th {color:#093; font-size:large; font-weight:bold;}
```

If we added those styles to the previous table, the result would look like Figure A1–4.

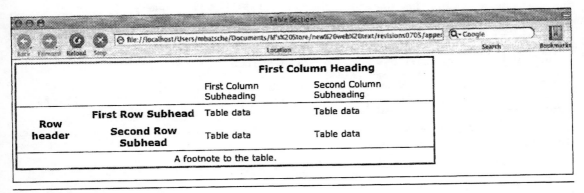

FIGURE A1–4   A table with styles

Appendix 2
Image Replacement
Techniques

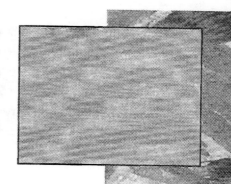

The use of CSS has done away with some older layout methods, such as using tables for layout grids. However, it has also provided new ways to implement some older techniques. For example, in older Web designs, designers would sometimes create graphic page banners so that they could use display fonts or fonts that were not loaded on most computers. The downside of this technique was that the images created accessibility problems since they couldn't be read with screen readers (unless the designers provided `alt` attributes). In addition, the graphic headers often were overlooked by search engines, and because they were graphics, they required a longer download than plain text.

With the advent of CSS, this practice declined in favor of using the CSS font properties. However, some designers have now developed a system that combines the old and the new techniques, placing graphic headers in the CSS so that the page code loads more quickly and (theoretically at least) providing searchable text that can be read by screen readers. The technique still has some serious drawbacks, which have led to some variations, but image replacement as a technique is still being explored and offers intriguing possibilities.

FAHRNER IMAGE REPLACEMENT

The original method of replacing text with an image, using the `background` property in CSS, was developed by Todd Fahrner and popularized by Douglas Bowman (2003). The process is relatively simple:

1. Create a stylish graphic heading, using your favorite display type in a graphic program like Photoshop or Fireworks, and make a note of the graphic's dimensions.

2. Create a heading with the text to be replaced.

```
<h1>My heading</h1>
```

3. Assign a unique id to the heading.

```
<h1 id="head1">My heading</h1>
```

4. Wrap a span around the heading content (you'll use it to hide the heading).

```
<h1 id="head1"><span>My heading</span></h1>
```

5. Hide the heading text by setting it to display:none on your style sheet.

```
#head1 span{display:none;}
```

6. Place your stylish graphic heading in the background of your heading; be sure to supply the width and the height of the graphic for the background dimensions.

```
#head1{width:529px; height:249px;
    background:url(head1.gif) #fff no-repeat;}
```

The image will now show through the invisible text you've supplied for heading one. The complete code looks like this:

```
<!DOCTYPE html PUBLIC "-//W3C//DTD XHTML 1.0
    Transitional//EN" "http://www.w3.org/TR/xhtml1/DTD/
    xhtml1-transitional.dtd">
<html xmlns="http://www.w3.org/1999/xhtml">
<head>
    <meta http-equiv="content-type" content="text/html;
charset=utf-8"/>
    <title>Fahrner Image Replacement</title>
    <style type="text/css">
#head1 span{display:none;}
#head1{width:529px; height:249px; background:url-
    (heading1.gif) no-repeat;}
</style>
</head>
<body>
<h1 id="head1"><span>My heading</span></h1>
</body>
</html>
```

In a browser it would look like Figure A2–1. You could use this graphic header as the banner on all of your pages, creating both unity and identification.

The advantages of this method are twofold. Since the image originates with CSS rather than HTML, browsers and devices that don't support CSS will show the text version given in the HTML. (The display will be normal since the display:none is a CSS property.) In addition, because the graphic is placed in the external CSS file rather than the HTML page, switching to a new graphic takes only one change rather than one on each page.

However, this method has several disadvantages as well. Although logically you might expect that a screen reader would ignore the display:none style rule

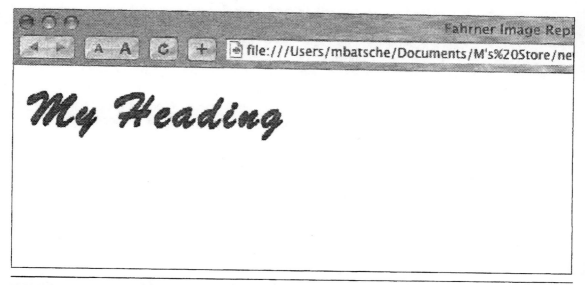

FIGURE A2–1   Fahrner image replacement

(since screen readers don't display text at all), in fact they obey the declaration. Thus, a user with assistive technology will be unable to see the graphic or hear the text version. In addition, the method adds extra markup with the span element, and if you have any users who have disabled their images but kept their CSS in force, neither the text nor the graphic will be visible.

LEAHY/LANGRIDGE IMAGE REPLACEMENT

A second image replacement method was developed by Seamus Leahy (2003) and Stuart Langridge (2003). The main difference between the Fahrner method and the Leahy/Langridge method is that the latter moves the text out of the way rather than using display:none. With this method you set the height of the heading to 0 and set padding-top to equal the height of the replacement image.

1. Create a stylish graphic heading, using your favorite display type in a graphic program like Photoshop or Fireworks, and make a note of the graphic's dimensions.
2. Create a heading with the text to be replaced.

   ```
   <h1>My heading</h1>
   ```
3. Assign a unique id to the heading.

   ```
   <h1 id="head1">My heading</h1>
   ```
4. Add the following rule to your stylesheet.

   ```
   #head1 {padding:46px 0 0 0; overflow:hidden; back-
   ground:url(heading1.gif) #fff no-repeat;
   height:0px;}
   ```

In this case the image is 46 pixels high, and the top padding is set to the same value, which will show the background graphic underneath. Setting the height to 0 will conceal the text for most browsers.

Unfortunately, Internet Explorer 5 for Windows has some problems with this method because of its misinterpretation of the box model. Working with Explorer 5 involves inserting special code to trick the browser into reading the correct dimensions. For an explanation of the solution to this problem, see Dan Cederholm's *Web Standards Solutions* (2004) or the Langridge or Leahy Web sites. The code will look like this:

```
<!DOCTYPE html PUBLIC "-//W3C//DTD XHTML 1.0
    Transitional//EN" "http://www.w3.org/TR/xhtml1/DTD/
    xhtml-transitional.dtd">
<html xmlns="http://www.w3.org/1999/xhtml">
<head>
    <meta http-equiv="content-type" content="text/html;
charset=utf-8" />
    <title>Leahy/Langridge Image Replacement</title>
    <style type="text/css">
#head1 {padding: 46px 0 0 0; overflow:hidden; back-
    ground:url(heading2.gif) #fff no-repeat; height:0px;}
</style>
</head>
<body>
<h1 id="head1">My heading</h1>
</body>
</html>
```

In a browser it would look like Figure A2–2 (I've used a different font in the graphic for variety).

There are a few problems remaining with the Leahy/Langridge method. It requires extra code to make Internet Explorer for Windows read the code properly,

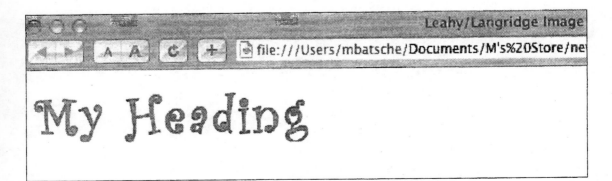

FIGURE A2–2   Leahy/Langridge image replacement

and again, any users who have their images disabled but not their CSS will see nothing at all. However, users with screen readers will hear the text version of the heading.

PHARK IMAGE REPLACEMENT

One other image replacement technique has been developed by Mike Rundle (2003; it's named for his Web site). Rundle uses a large negative text-indent value to hide the text, moving it so far off the page that it cannot be seen. To use the Phark technique, follow these steps:

1. Create a stylish graphic heading, using your favorite display type in a graphic program like Photoshop or Fireworks, and make a note of the graphic's dimensions.
2. Create a heading with the text to be replaced.

   ```
   <h1>My heading</h1>
   ```

3. Assign a unique id to the heading.

   ```
   <h1 id="head1">My heading</h1>
   ```

4. Add the following rule to your style sheet:

   ```
   #head1 {height:83px; text-indent:-5000px; back-
   ground:url(heading3.gif) #fff no-repeat;}
   ```

Rundle essentially pushes the text out of the way so that it's unseen by the user but available to text readers. The code looks like this:

```
<!DOCTYPE html PUBLIC "-//W3C//DTD XHTML 1.0
    Transitional//EN" "http://www.w3.org/TR/xhtml1/DTD/
    xhtml1-transitional.dtd">
<html xmlns="http://www.w3.org/1999/xhtml">
<head>
    <meta http-equiv="content-type" content="text/html;
charset=utf-8" />
    <title>Phark Image Replacement</title>
    <style type="text/css">
#head1 {height:83px; text-indent:-5000px;
    background:url(heading3.gif) #fff no-repeat;}
</style>
</head>
<body>
<h1 id="head1">My heading</h1>
</body>
</html>
```

In a browser it would look like Figure A2–3 (once again, I've changed the font).

Although Phark takes care of the problems with screen readers and Internet Explorer 5 for Windows, it still doesn't solve the images-off/ CSS-on problem. Granted, few users surf the Web this way, but anyone who does won't be able to see your heading.

FIGURE A2–3 Phark image replacement

Image replacement is an interesting technique that's being widely used and experimented with by Web designers. It's possible that some designer will come up with a new method that takes care of the one remaining problem shared by all three of these methods. You can keep up-to-date with new developments by consulting Dave Shea's ongoing coverage of image replacement at http://www.mezzoblue.com/tests/revised-image-replacement/.

SOURCES

Bowman, D. 2003. Using background-image to replace text. http:/ /www.stopdesign.com/also/articles/replace_text (accessed June 14, 2005).

Cederholm, D. 2004. *Web standards solutions.* New York: Friends of Ed.

Langridge, S. 2003. A new image replacement technique. http://www.kryogenix.org/code/browser/lir/(accessed June 14, 2005).

Leahy, S. 2003. Image replacement—No span. http://moronicbajebus.com/(accessed August 14, 2005).

Rundle, M. 2003. Accessible image replacement. http://phark.typepad.com/phark/2003/08/accessible_imag.html(accessed June 14, 2005).

Appendix 3
Print Style Sheets

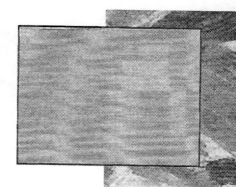

Using CSS, you can create style sheets that will be used only for printed documents. This means you can create a separate style sheet for your screen layout and for your print layout, removing extraneous pieces that are unnecessary for printed pages, like sidebar menus or banners. You can also write a style sheet using more familiar print styles, like points for font sizes and inches for margins.

SETTING UP A PRINT STYLE SHEET

To create your print style sheet, you first need to designate the kind of style sheet it is. You do this by adding a media attribute to your link tag. There are several possible values for media, but three are used most frequently:

- ❑ screen—A screen style sheet applies to the layout to be used on the computer screen.
- ❑ print—A print style sheet applies to the layout in the printed version of the page.
- ❑ all—These style sheets apply to both print and screen; all is the default value for media.

The link tag will look like this:

```
<link rel="stylesheet" media="print" type="text/css"
    href="print.css">
```

Thus, you'll have two style sheets: one for your screen layout and another for your print layout. The page head section will look something like this:

```
<!DOCTYPE html PUBLIC "-//W3C//DTD XHTML 1.0
    Transitional//EN" "http://www.w3.org/TR/xhtml1/DTD/
    xhtml1-transitional.dtd">
<html xmlns="http://www.w3.org/1999/xhtml">
<head>
```

```
    <meta http-equiv="content-type" content="text/html;
charset=utf-8"/>
    <title>Using Print Styles</title>
    <link rel="stylesheet" href="styles.css"
type="text/css" media="screen"/>
    <link rel="stylesheet" href="print.css"
type="text/css" media="print"/>
</head>
```

USING PRINT STYLES

Several properties can be used on your print style sheets to make your pages easier to print. First of all, you'll use different font sizes, and you may also use different font families. Some readers prefer serif fonts for printed documents (e.g., Times, Palatino, Century Schoolbook) and sans serif fonts for screens. You can set the font size in points, which is the usual unit of measure for printed fonts. And you can set the color to black on white, the most legible combination for printed pages.

If you have sidebars, banners, or menus, you can remove them from the print style layout so that you're including only the main content, or you can add supplemental information like sidebars at the end of the main content. You can also remove any background images you use on the screen layout, including those you've embedded for image replacement (see appendix 2). In fact, if you're using image replacement for your screen layout, you'll need to remove all the associated styles for print to make sure that your banner/title is printed.

Finally, you can adjust margins to typical print measurements, frequently one inch at top and bottom and one and a quarter inches at left and right.

TESTING PRINT STYLES

Browser support for print style sheets is good, but there are some quirks to be aware of. Different browsers may treat point sizes differently, for example. You'll want to see for yourself what happens when you print in different browsers. One way to do this without printing lots of test sheets is to use the Print Preview option (usually in the File menu or the Print controller).

Print style sheets are applied seamlessly; that is, users won't realize that another style sheet has been applied until they see the printed page. However, you may want to alert your users to the difference in layout. You can do this by setting up a link to a page with the print layout applied. You can even set up a separate print version for multipage documents, particularly since many users are accustomed to looking for an option for a print layout. However, the print styles will be applied without a separate page having been created.

PRINT-SPECIFIC CSS

CSS includes some properties that are expressly designed to help with printing. Unfortunately, support for these options is mixed, so be sure to test them in various browsers before relying on them.

Page Breaks

When you have a page that stretches down several screens, it will be split up into multiple pages when it's printed. Using CSS, you can set the page breaks so that they occur where you want them to occur. There are two properties involved: `page-break-before` and `page-break-after`. The values are the same:

❑ `always`—The page break will always occur either before or after the element.
❑ `avoid`—The page breaks will occur before or after the element only when absolutely necessary.
❑ `auto`—The browser will decide where the page break occurs; `auto` is the default.

You can also use a third property to keep an element from being divided between two pages: `page-break-inside: avoid`. The property looks like this:

```
h1{page-break-after:avoid;}
```

Widows and Orphans

A widow is a single line of a longer paragraph, displayed alone at the beginning of a page; an orphan is a single line, displayed alone at the end of a page. Both are traditionally avoided in printing. CSS allows you to designate how many lines of an element can appear alone. There are two properties: `widows` and `orphans`. The property looks like this:

```
p{widows:2; orphans: 2;}
```

Here I'm specifying that my paragraphs should have a minimum of two lines at the top and the bottom of the page.

Appendix 4
Blogging

Blog, short for *Weblog*, is a Web site that posts articles or links of interest on a particular topic, frequently in reverse chronological order. Early Weblogs were created using HTML and uploaded by FTP (file transfer protocol), but now automated blog tools exist to create and maintain such sites. Consequently, blogging has become a worldwide phenomenon. In fact, Cory Doctorow et al. (2002) state that 1,500 to 3,000 new bloggers log on every day (ix). The estimated total number of blogs varies. The National Institute for Technology and Liberal Education has a blog census, which has indexed over 250,000 blogs, but it has had over 1,000,000 blog URLs submitted that have not yet been authenticated (http://www.nitle.org/).

Subjects for blogs run the gamut from diaries to corporate media campaigns. They can be written by individuals or by collaborative teams, and visitors can leave their own comments on some blogs, creating a community of readers who may then go on to create blogs of their own on similar or related topics. Some blogs are little more than lists of hyperlinks, whereas others provide article summaries or complete articles and sometimes allow comments, articles, and ratings from readers.

Blog software programs are actually content management systems (see chapter 13), providing tools for composing and posting blog content, as well as linking to other blogs. Most blog software archives older blog entries and provides a static, or unchanging, URL for them; this static link is referred to as a permalink. Many blog programs also create RSS (rich site summary) feeds for blog entries. RSS is an XML application that reads blog entries and then provides users with a list of headlines from various blogs and news sites, along with hyperlinks and summaries. These headlines, in turn, can be read with programs called feed readers or news readers.

BLOGGING HISTORY

Blogs are a logical outgrowth of other types of digital communities, including Usenet (a long-standing newsgroup service), e-mail lists, and electronic bulletin boards. In fact, some of the terms widely used with blogs come from these earlier communities—for example, *threads* (i.e., a number of messages on a particular topic, a term borrowed from e-mail lists) and to *post* (i.e., to upload a message, a term taken from electronic bulletin boards).

The term *Weblog* was coined by Jorn Barger in 1997; the shorter term *blog* (from "we blog") came from Peter Merholz in 1999. The first blogging tool was the Blogger program, developed by Evan Williams and Mego Hourihan for Pyra Labs in 1999 and purchased by Google in 2004. The development of RSS by Dave Winer, along with blog-reading utilities, also helped to increase blog popularity. The term *Weblog* entered the Oxford English Dictionary in 2003.

CREATING A BLOG

The earliest forms of Weblogs were created with HTML and text and then uploaded using FTP. However, the development of blogging software has made the process of creating and updating log entries considerably simpler. We'll discuss some of the alternatives in blogging software, but first, here's an overview of how you might create a blog entry.

Making an Entry

A typical blog entry has several parts. First, you create a title that indicates the subject of the entry. The title serves as a kind of headline for the post, summing up the subject of the entry in an interesting, attention-getting way.

Some blogging software allows you to create both short and extended versions of entries. The shorter, more condensed version can be scanned quickly and can serve as a kind of abstract for the longer version. The shorter version can also be sent to regular readers of your blog via e-mail as a way to build interest in your blog.

Both short and long entries can be made available for comments or closed to comments. In most cases you can edit any comments you receive or delete them if they're inappropriate. All blog entries should also be dated so that users know how current the information is.

Some blogging software allows you to initiate and accept pings, that is, tests of whether particular hosts are operating properly. With pings the blogging software alerts other blogs to updates in your entries. You can provide URLs of other blogs that cover the same topics, and some blogging software keeps track of URLs that you've pinged in the past. Frequently, you can also look at other blogs that have pinged your entry and delete the link if you don't want your entry linked to theirs. Or you can send an alert concerning the entry to a group of readers, usually people who have commented on the blog in the past or people who are interested in the topic of the blog.

Although the options provided by software vary, all blogging software has ways for you to create titles and entries, and most software allows you to manage any comments you receive. Most programs also provide a means of pinging and linking to other blogs. You can choose the blogging software you want to use, based on your needs and your access to a Web server.

BLOGGING SOFTWARE

All blogs must be hosted on a Web server somewhere. Some programs can be installed on your own Web server, assuming you have access to one. Others publish your blog on their own sites by allowing you to upload your blog files via FTP. Programs that must be loaded on your own server include Movable Type, Blosxom, Greymatter, Slash, and Zope. Programs in which you upload your blog to another site include Blogger, Live Journal, and Radio UserLand. Social networking Web sites such as Myspace, Friendster, and LiveJournal also provide on-site blogging software for your use. Installing the blogging software on your own server is convenient, but the installation and maintenanace require some background in Web administration. Having your blog on someone else's server requires somewhat less technical expertise, but these programs also have their drawbacks. Some free services will put up banner ads on your blog, and FTP uploads can be time-consuming.

Some blogging software is free (e.g., Slash, Zope, Greymatter, and Blosxom), whereas some programs have both free and subscription versions (e.g., LiveJournal and Blogger). Some subscription services are free for personal use (e.g., Movable Type); others offer trial versions (e.g., Radio UserLand). Sometimes the blogging software itself is free, but you pay an additional fee to produce or receive RSS feeds. Choose your blogging software according to your level of technical expertise and your decision about paying for additional services.

PERSONALIZING YOUR BLOG

All blogging software comes with default templates and sometimes with other optional templates as well. Most allow you to edit the templates to some degree, changing things like colors and type fonts. The code for the templates may be written in a combination of HTML and XML, but the HTML will be editable.

If you're adding a blog to an existing Web site, you'll probably want to reconfigure the template so that it looks like the other pages on the site. You can copy the page template and then paste it into an HTML editing tool like Dreamweaver or BBEdit. Because the template is usually made up of a combination of HTML and CSS, you can take out the CSS classes and ids and substitute your own. But you'll want to be careful to maintain the tags that are specific to the blog, that is, those that call up and place the blog's content. Once you know which tags these are, you can simply insert them into your Web site's page template.

SOURCES

Doctorow, C., R. Dornfest, J. S. Johnson, S. Powers, B. Trott, and M. C. Trott. 2002. *Essential blogging*. Sebastapol, CA: O'Reilly.

Appendix 5
Character Entity
References

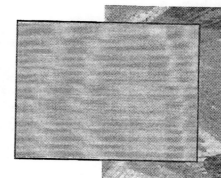

Character entity references, or character references, are a means of inserting symbols and characters into your HTML and XHTML that aren't part of your text's normal encoding. Instead of typing in the symbol from your keyboard (which would result in garbled characters online), you can insert a set of numbers or letters that the browser will translate into the symbol. Listing the entire Unicode character set is far beyond the scope of this appendix, but you can visit the Unicode Web site at http://www.unicode.org/ or the W3C at http://www.w3.org/ for a more complete listing.

The format of the character references is fairly consistent: all of them begin with an ampersand (&) and end with a semicolon (;) Between the ampersand and the semicolon, you may have a hexadecimal number, which begins with #x; a regular number, which begins with #; or a named reference, which uses the character name (e.g., &, which is the character reference for an ampersand).

In XHTML, code using character references looks like this:

```
<p>Gabriel Jos&eacute; Garc&iacute;a M&aacute;rquez, a
Colombian author and journalist, was the winner of the
1982 Nobel Prize for Literature.</p>
```

In a browser it would look like Figure A5-1.

Following is a list of some of the character entity references that are commonly used in XHTML, not including the long list of mathematical and technical symbols. Again, you can find a more complete listing at the Unicode site or the W3C site.

Gabriel José García Márquez, a Colombian author and journalist, was the winner of the 1982 Nobel Prize for Literature.

FIGURE A5-1 Character references

SYMBOL	DESCRIPTION	ENTITY NAME	ENTITY NUMBER
"	Quotation mark	&quot;	&#34;
'	Apostrophe	&apos;	&#39;
&	Ampersand	&amp;	&#38;
<	Less than	&lt;	&#60;
>	Greater than	&gt;	&#62;
	Nonbreaking space		
¡	Inverted exclamation point	&iexcl;	&#161;
©	Copyright	&copy;	&#169;
®	Registered trademark	&reg;	&#174;
™	Trademark	&trade;	&#8482;
°	Degree	&deg;	&#176;
¶	Paragraph	&para;	&#182;
¿	Inverted question mark	&iquest;	&#191;
À	A, grave accent	&Agrave;	&#192;
Á	A, acute accent	&Aacute;	&#193;
Â	A, circumflex	&Acirc;	&#194;
Ã	A, tilde	&Atilde;	&#195;
Ä	A, umlaut	&Auml;	&#196;
Å	A, ring	&Aring;	&#197;
Æ	Æ	&AElig;	&#198;
Ç	C, cedilla	&Ccedil;	&#199;
È	E, grave accent	&Egrave;	&#200;
É	E, acute accent	&Eacute;	&#201;
Ê	E, circumflex	&Ecirc;	&#202;
Ë	E, umlaut	&Euml;	&#203;
Ì	I, grave accent	&Igrave;	&#204;
Í	I, acute accent	&Iacute;	&#205;
Î	I, circumflex	&Icirc;	&#206;
Ï	I, umlaut	&Iuml;	&#207;
Ñ	N, tilde	&Ntilde;	&#209;
Ò	O, grave accent	&Ograve;	&#210;
Ó	O, acute accent	&Oacute;	&#211;
Ô	O, circumflex	&Ocirc;	&#212;
Õ	O, tilde	&Otilde;	&#213;
Ö	O, umlaut	&Ouml;	&#214;

Ø	O, slash	&Oslash;	&#216;
Ù	U, grave accent	&Ugrave;	&#217;
Ú	U, acute accent	&Uacute;	&#218;
Û	U, circumflex	&Ucirc;	&#219;
Ü	U, umlaut	&Uuml;	&#220;
à	a, grave accent	&agrave;	&#224;
á	a, acute accent	&aacute;	&#225;
â	a, circumflex	&acirc;	&#226;
ã	a, tilde	&atilde;	&#227;
ä	a, umlaut	&auml;	&#228;
å	a, ring	&aring;	&#229;
æ	æ	&aelig;	&#230;
ç	c, cedilla	&ccedil;	&#231;
è	e, grave accent	&egrave;	&#232;
é	e, acute accent	&eacute;	&#233;
ê	e, circumflex	&ecirc;	&#234;
ë	e, umlaut	&euml;	&#235;
ì	i, grave accent	&igrave;	&#236;
í	i, acute accent	&iacute;	&#237;
î	i, circumflex	&icirc;	&#238;
ï	i, umlaut	&iuml;	&#239;
ñ	n, tilde	&ntilde;	&#241;
ò	o, grave accent	&ograve;	&#242;
ó	o, acute accent	&oacute;	&#243;
ô	o, circumflex	&ocirc;	&#244;
õ	o, tilde	&otilde;	&#245;
ö	o, umlaut	&ouml;	&#246;
ø	o, slash	&oslash;	&#248;
ù	u, grave accent	&ugrave;	&#249;
ú	u, acute accent	&uacute;	&#250;
û	u, circumflex	&ucirc;	&#251;
ü	u, umlaut	&uuml;	&#252;
–	en dash	&ndash;	&#8211;
—	em dash	&mdash;	&#8212;
'	left single quotation mark	&lsquo;	&#8216;
'	right single quotation mark	&rsquo;	&#8217;
"	left double quotation mark	&ldquo;	&#8220;
"	right double quotation mark	&rdquo;	&#8221;
. . .	horizontal ellipsis	&hellip;	&#8230;
€	euro	&euro;	&#8364;
¢	cent	&cent;	&#162;
£	pound	&pound;	&#163;
¥	yen	&yen;	&#165;

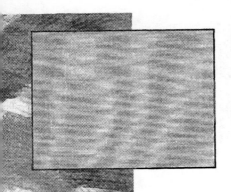

Appendix 6
Browser-Specific Tags
and Deprecated Tags

Browser-specific tags have been created by browser manufacturers to be used in particular browsers; they cannot be applied in any browser other than the one for which they were developed. Deprecated tags have been superseded by other tags or CSS properties. All of these tags should be avoided in your designs.

TAGS AND ATTRIBUTES FOR INTERNET EXPLORER ONLY

ELEMENT	FUNCTION	EFFECT
nowrap	Attribute	Prevents text from wrapping
action	Attribute	Narrows search terms
marquee	Tag	Creates scrolling text
bgsound	Tag	Inserts background sound
color (with HR)	Attribute	Colors horizontal rule
comment	Tag	Places comments in code
controls	Attribute	Adds controls with in-line movie
dynsrc	Attribute	References AVI movie for in-line display
loop	Attribute	Repeats movie for stated number of times
start	Attribute	Delays movie playback
leftmargin	Attribute	Indents left margin relative to left edge of browser window
topmargin	Attribute	Sets margin of space at top of document
notab	Attribute	Takes area out of tabbing sequence on page
bordercolor	Attribute	Sets color for table borders
bordercolorlight	Attribute	Sets color for table borders
bordercolordark	Attribute	Sets color for table borders

TAGS AND ATTRIBUTES FOR NETSCAPE ONLY

ELEMENT	FUNCTION	EFFECT
blink	Tag	Makes text blink on and off
keygen	Tag	Generates public key to encrypt HTML forms
layer	Tag	Allows use of layers on a page
lowsrc	Attribute	Allocates space for image as document is rendered
marginheight	Attribute	Defines top margin of browser window
marginwidth	Attribute	Defines left margin of browser window
multicol	Tag	Displays text in multiple columns
spacer	Tag	Inserts nonbreaking space of specified size

DEPRECATED TAGS

INSTEAD OF THIS	USE THIS
BASEFONT	Font properties in CSS
BIG	font-size:bigger (CSS)
CENTER	text-align:center (CSS)
DIR (directory list)	ul
FONT	Font properties in CSS
MENU	ul or ol
S or STRIKE	text-decoration: line-through (CSS)
SMALL	font-size:smaller (CSS)
U	text-decoration: underline(CSS)

DEPRECATED ATTRIBUTES

INSTEAD OF THIS	USE THIS CSS PROPERTY
align	text-align or vertical-align (text) or float (graphics)
alink	a:active (a pseudoclass)
background	background-image
bgcolor	background-color
border (img)	border
cellpadding	padding
cellspacing	margin
clear	clear
color	color
compact	margin or padding

Continued

Continued

INSTEAD OF THIS	USE THIS CSS PROPERTY
face	font-family
height (td and th)	height
hspace	padding or margin
link	a:link (a pseudoclass)
noshade (hr)	border-color or color
size	font-size
text	color
type (ol, ul, li)	list-style
valign	vertical-align
vlink	a:visited (a pseudo class)
vspace	margin or padding
width (hr, td, th)	width

Index

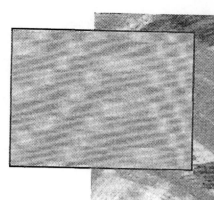